An Alzheimer's Surprise Party

Unveiling the Mystery, Inner Experience, and Gifts of Dementia

"I strongly recommend this as an original method of understanding and dealing with people in Alzheimer's and other dementias."

DR. ARNOLD MINDELL, PhD
Jungian Training Analyst, founder of Process Oriented Psychology

"This book is revolutionary in providing hope and meaning for patients, relatives, friends, caregivers, and professionals in the management of Alzheimer's and other dementias."

DR. ROSEMARY SHINKWIN, MD, MRCPsych
Consultant Psychiatrist, Mercy University Hospital, Cork, Ireland

ALZHEIMER'S ASSOCIATION RECOMMENDED RESOURCE

An Alzheimer's Surprise Party

Unveiling the Mystery, Inner Experience, and Gifts of Dementia

Tom Richards

Stan Tomandl

ISBN 978-0-578-02276-5 ~ Second Edition ~ Paperback

ISBN 978-1-4116-7456-1 ~ First Edition ~ Hardcover

Published by:

Interactive Media

1218 Roosevelt Avenue

Glenview, Illinois

60025 USA

Cover design by Donna Brady with Cindy Trawinski

Second Edition, third printing

First printing 1995

To our elders

Table of Contents ~

In Appreciation

Thanks to Stanley's family and friends and the staff at Illinois Veterans Home, Manteno, especially Thyra for all the love and support and prayers. Thank you to our mentors, supervisors, teachers, colleagues, loved ones, and supporters of this project. Thanks to Julie Diamond; Joe Goodbread; Ann Jacob; Olga La Luz; Bob Middleton; Amy Mindell; Arny Mindell; Max Schupbach; and Peter Thomas: for helping us process physical, emotional, and spiritual needs around Stanley's Alzheimer's and his death.

Thanks to Stanley and his wonderful spirit. He always exclaimed, "Death has got to be the greatest adventure!"

Special appreciation to Drs. Arnold and Amy Mindell for their insight and courage to bring forward to the world, the art of communication with people in faraway states of altered consciousness.

Preface

In this groundbreaking work Tom Richards and Stan Tomandl offer a new and exciting shift in the therapeutic mindset for family, friends, and professionals working with people in Alzheimer's and other dementia states.

They propose that people with Alzheimer's dementia are not merely spiraling downward into "mindless pathology", but are human beings in states of altered consciousness, parallel realities that may be important and meaningful experiences for them, their families, and for society.

The authors liken the Alzheimer's dementia patient to a hero or heroine on a mythic journey venturing into the unknown, encountering extraordinary and sometimes divine experiences. They show by means of one such detailed case history that this journey can reveal processes involving: completion of "unfinished business", such as resolving individual and family issues; "harvesting", such as recalling and savoring life experiences; "imparting blessings", such as openly accepting loved ones; creating "sacred space" by tapping into a sense of something larger than ourselves; garnering "meaning" by exploring formative experiences and essential beliefs; and making "spiritual connections" like immersing in the beauty of eternity.

Along with verbatim bedside reports, comprehensive summaries, theoretical discussions, practical exercises, plus an extensive index and bibliography, Stan and Tom demonstrate techniques for communicating with individuals in extreme and altered states of consciousness, including advanced Alzheimer's dementia, delirium, and coma. These techniques are based on the sentient communication and facilitation skills of Process Work, learned from its innovators Drs. Amy and Arnold Mindell.

In showing that an individual, even in an advanced Alzheimer's dementia state, is able to communicate with others, the authors demonstrate that such patients can often make conscious rational decisions, thus adding a significant new dimension to ethical and legal debates around altered consciousness and end of life conditions.

This is a truly informative book and a must for patients and families affected by Alzheimer's dementia, and the professionals working with them.

Dr. Rosemary Shinkwin, MD, MRCPsych

Poem

Do you remember the ice age?

. . . . .

When death had all dominion

Though we had no name for it

No numbers to crunch, no memory

Once again

We've assembled for the wrong play

Morris Wattenberg, 1994

Chapter One

An Alzheimer's Surprise Party

Before beginning our story

Alzheimer's dementia is one of the greatest and most intriguing mysteries of our time. It is a pandemic that has, to a significant degree, defied explanation, prevention, and cure for over a hundred years. This book is an invitation to open a fresh view on this subject that panics most people when personally confronted with it. We present a deep, gentle, empathetic approach to further understanding this mystery.

Between the covers of this book is the passionate story of succeeding beyond conventional wisdom, to stay in communication with, in relationship to, and in love with a husband, a father, and a friend who goes in and out of very remote states of consciousness. As our story unfolds we introduce new Process Work ideas and interventions and provide training exercises for relating to people with Alzheimer's and other dementias. These ideas can also help lessen the stigma of dementia and relieve suffering by remaining open to the inner worlds of dreaming and spirituality during the elder phase of people's lives.

Using the sentient communication and facilitation skills of Process Work, learned from its innovators Amy and Arnold Mindell, we open our story with an Alzheimer's surprise party! We close nine chapters later with a proposed psychosocial-spiritual prescription for helping with Alzheimer's dementia, and an invitation to continue exploring the mystery.

The essence of our story is, of course, love . . .

It begins with the sunburst. As I unfold the colorful party decoration, Fran's eyes light up and the party comes to life. Her seventy-eight year old husband Stanley has been in the infirmary of the Illinois Veterans Home in Manteno, Illinois, for the past seven days under a death vigil. He is dying of complications after a three and a half year bout with advanced Alzheimer's dementia. Stanley is now at the end stage, slipping in and out of semicomatose states. And we are throwing him a surprise party? It had been the furthest thought from Fran's mind. Her prospects for the day had been dismal. She had expected to repeat her visit of three days ago, when she sat watch by her unresponsive husband and helplessly listened to his labored breathing.

Fran: *Yes, a party! Stanley, we are going to have a party!* Stanley startles her by opening his half closed unfocused eyes a little wider and nodding his head slightly. These are minimal signals, very small movements from a physical perspective, but very significant positive feedback signals from a man deep in altered consciousness, close to a coma state. The three of us all agree then. All right! It is party time!

So I put up the sunburst. I have tape with me for the occasion, but decide to hang it on the IV stand. It covers up the medical paraphernalia and makes a silent statement on behalf of the family's wish for no further heroic medical measures.

Fran: *Your sweetheart's here, and I love you. Oh, you're raising your eyebrows! It's too bad we didn't stop to get some wine.*

I joyously surprise them as I reach in my bag: *Not to worry. I have white zinfandel (Fran and Stanley's favorite) and champagne!*

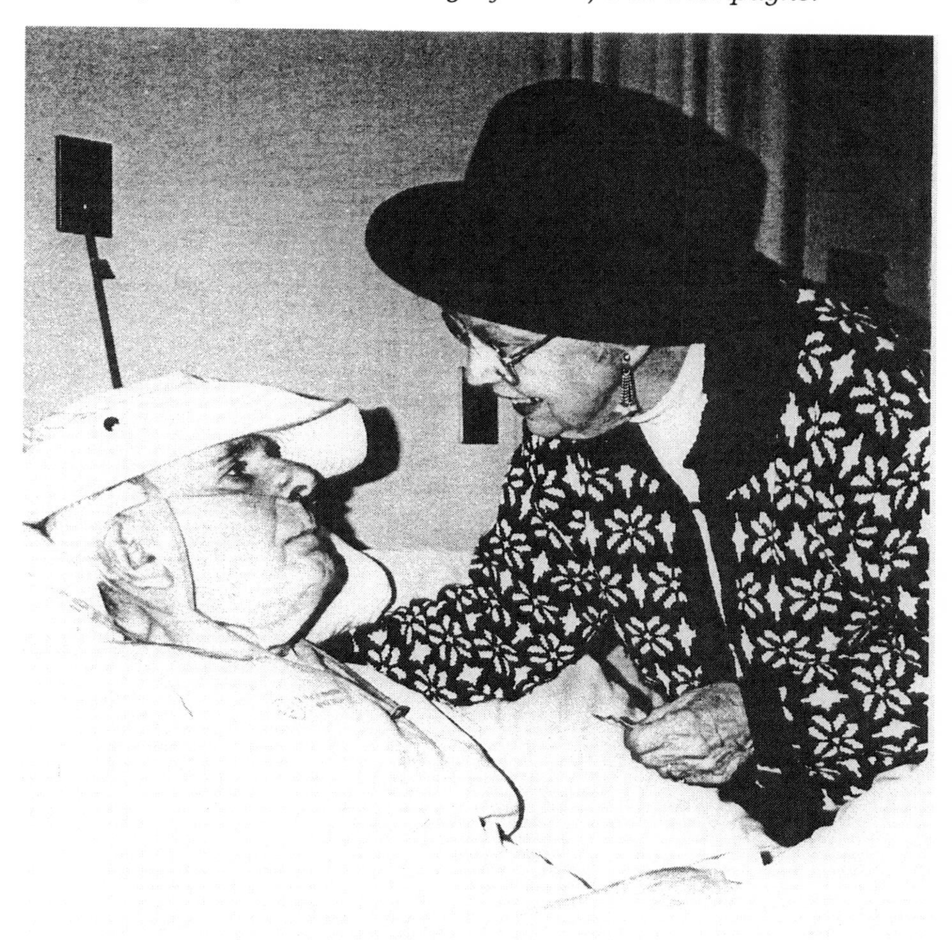

Stanley Tomandl, end stage Alzheimer's patient, completely out of his semicomatose state during his death vigil, relating to wife Francis while enjoying new hats, wine, champagne, ice cream, music, songs, hymns, prayers, intimacy and sacred ceremonial closure at their "Alzheimer's Surprise Party" ~ the day before Stanley dies after three and a half years of advanced Alzheimer's and seven days in a metabolic coma.

Fran: *Stanley, you love parties, don't you.* Stanley immediately coughs, indicating more positive feedback. *Do you want some ice cream?*

Tom: *You've been waiting for your party, haven't you!* Fran and I had stopped at the commissary for double scoops of ice cream. Fran starts feeding Stanley ice cream while I pull out a bag of hats. Stanley has always been a hat man and even wore one in his basement workshop.

Fran: *Oh, hats! He loves hats! What kind of hat do you want today, Stanley?* Fran and I enthusiastically try all the hats on Stanley and each other. We decide on an Australian bush hat for Stanley. Fran says: *It looks just like Stasche's style* (their oldest son, also a "Stanley" or "Stan", nicknamed" Stasche", and co author of this book*).* She dons a stunning black fedora, and I put on a traditional baseball cap.

Fran: *Boy, you look great!* Stanley coughs again. Fran is now in full swing. *Let's get some pictures. What should I do, kiss him?*

Tom: *You can do whatever you want to do. I will take a few pictures.*

Fran gives him a couple of big smackers and exclaims: *I kissed you, Stanley. ~ Look at him look at me! ~ I kissed you honey. Yeah, isn't that sweet. You're a doll!*

Tom: *We're going to write a book about you, Stanley, remember.*

Stanley starts making short sounds: *Uhg . . . Uhg . . .*

I repeat his sounds, making them slightly louder and longer: *Uuhhgg . . . Uuuhhhgggg . . .* This gives Stanley feedback as to what his voice sounds like and helps support him to go farther and modify his vocalizations if he wants to and is able.

Fran: *You're breathing hard. I hope we don't get you too excited.* Fran takes some pictures of Stanley and me and says: *Stasche will like to see you having a party! ~ He sees and everything!*

Tom: *Yes, he looks great. He is completely out of his coma.*

Fran: *We are celebrating lots of stuff.*

Tom: *He is the only guy I know to attend his own going away party.*

Fran: *I think it's never happened before.* Stanley's body is on the last stage of a one-way journey. Fran remembers how Stanley had often told his children that death had to be the "greatest adventure" of all. A celebration seems a fitting send-off for his adventure.

Tom: *You are getting very flushed, Stanley. We are going to celebrate your life.* Fran and I call him by all his names and nicknames; Stanley, Stan, Stanislaus, Stump, and Grandpa.

Now, room decorated, with all of us dressed to party, we break out the champagne. Fran dips her finger in the champagne and touches it to Stanley's lips. He licks his lips, so we all toast his life as he swigs champagne from a paper cup.

Fran: *He is breathing kind of heavy. We excited him too much.* Fran's concern has come up three times now. Stanley is dying, why not let his excitement come up to a certain extent?

Stanley coughs and starts "talking": *Hmmm . . . Hmmmm . . .*

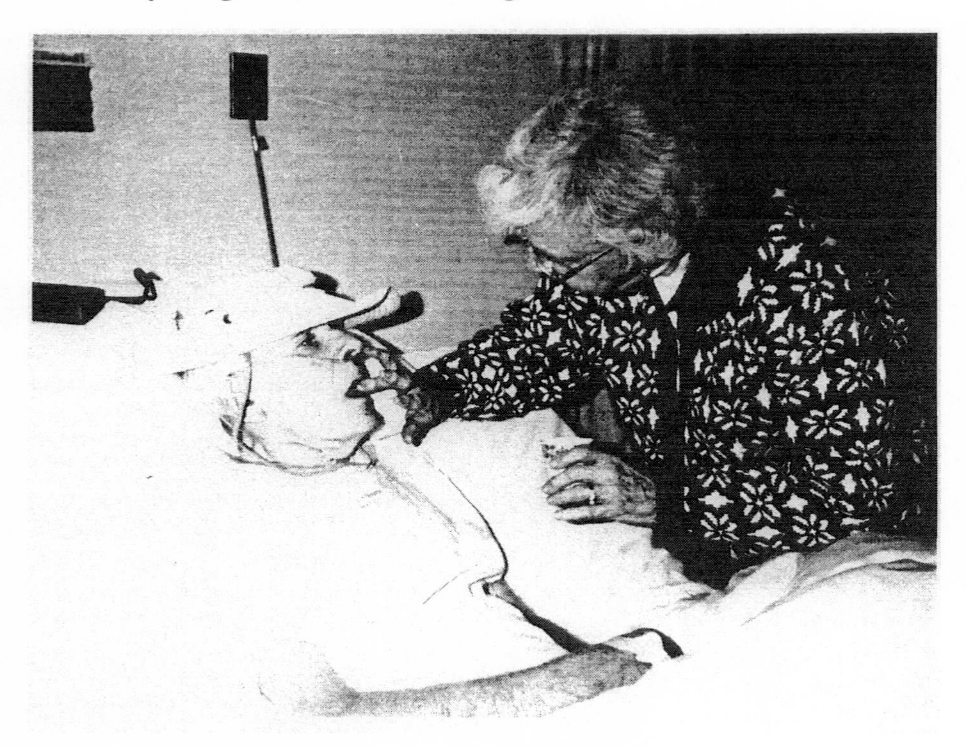

Chaplain ~ Lord's Prayer

At this point I realize it is time for my "chaplain" role and ask Fran her wishes for anything to be included in the celebration. She interprets it as a request being made to the master of ceremonies and defers to Stanley: *He is always the master of ceremonies.*

I clarify my request: *Are there any last rites you want included?*

Fran starts the last rites by reading a newspaper clipping entitled "What Money Can (and Can't) Buy." Between each phrase she hugs and kisses him: ~ *He's looking at me . . . awww. He always said a prayer at lunchtime and we always said the Lord's Prayer together.*

I suggest: *Maybe you'd like to do that now.* We recite an emotional rendition of the Lord's Prayer together and make another champagne toast: *Let's drink to Stanley's life.* Then we take some more pictures.

Fran: *I love you forever. You call me tweedy and I call you tweedy. We're tweedies.*

Stanley coughs; the immediacy and power of his coughing reaction indicates strong positive feedback.

Tom: *He likes that hat. ~ This hat has had a life, so I am giving you this hat. This hat is yours to keep.*

Fran: ~ *Look at him raise his eyebrows.* More positive feedback.

Tom: *Well, this is what your son Stasche said he would do if he was here. He'd throw you a party. So we're throwing you a party for Stasche. He's thinking about you. He said he'd throw you a wake. He'd throw you one of those wakes like you attended when you were a mortician.*

Is it okay to take your picture? It is going to be on the cover of your book, you know. ~ Fran, did you see the feedback? He moved his legs!

Master of ceremonies

Fran: *He used to call bingo and cheer everybody up at Brandel Care Center. And he used to call square dances. He loved to be the master of ceremonies.*

You know your sister Marie. She was bossy, and she told you what to do just before your were going to do it, and you didn't like that very much.

Tom: *Well he's broadcasting on all channels. From his breathing signals* (breathing into his chest, diaphragm, and lower stomach all at the same time), *he is seeing, listening, feeling, and moving. He's on a trip here, a bon voyage trip.* Fran gives him a big gulp of bubbly champagne. I check in with Fran and ask how she is doing.

"In the Mood"

Fran: *I'm okay.*

We're having a party for you, Stanley. Do you know how old you are? You are seventy-eight years old. Kids ask me about how do you know you should get married? I say it is just a natural thing. We just figured we belonged together. We belong together don't we? There is no question about what we did. We could dance and sing to "Elmer's Tune" or "In the Mood." Fran sings the tune.

Stanley coughs and sings: *Oh . . . Ahh . . .*

I cheerlead: *Stanley, you're doing great. You're doing really good. Do whatever you have to do.*

Fran quips: *How about a game of cribbage or gin rummy?* This was a near nightly pre romance ritual in front of their fireplace for years.

Tom: ~ *It looks like Stanley is going inside.* At this point I am calling attention to the change in Stanley's signals; eyes closing, stillness in his body, slower deeper breathing, and paler skin tone indicating a deep inner sentient state. Advanced Alzheimer's patients are extremely sensitive and extremely powerful, at least in one way. If they need a break or don't like what is going on, they can stay inside or go farther away.

Fran asks: *Are you leaving us? You're going to go to a different part of the world, huh? I'll see you there sometime. Okay? Jeanne (daughter) and Susan (granddaughter) and Bob and Rita (friends) were all here to see you last weekend.*

Stanley grows very animated: *Ahhhug . . . Ahhhug . . . Ahhhug . . .*

I instruct Fran: *Fran, guess into what he is trying to say.* Fran may be able to intuit what he is trying to say, and then Stanley may be able to confirm her guesses with sounds or movements. Stanley may be clearly saying words in his mind, but he has lost the ability to clearly form the words in his throat and mouth. He may not be able to connect his inside speech to what he is saying out loud.

So I encourage his feedback loop between inner and outer speech. I repeat his phrases and extend them: *Ahhhhhhugu . . . Ahhhhhhuguhhh . . . This is a great trip. A great ride. Pretty exciting. Yea, an exciting ride!*

Fran: *I wish you could tell us about it. I know you said it was beautiful!*

Tom: *We got close to eternity there a couple times.* "Beautiful" and "eternity" refer to words that Stanley uttered at a session nine months previous. He had pneumonia and lay very close to death in a metabolic coma.

Fran: *Yep, a beautiful trip.* We break out the wine and toast his great trip.

Music

Tom: *I learned a new song on my guitar for you, Stanley. My guitar is out in the car and I can get it if you are interested.*

Fran: *Oh yes, Stanley loves music. He loved to play the tuba.* I go out to the car. I leave the tape recorder on. While I'm gone Fran says: *Everybody loves you, Stanley. I love you. Your children love you. Your grandchildren love you. Your great grandchildren love you. Your friends love you. It's nice to be loved by so many people.* From somewhere in the background soft romantic music mysteriously drifts into the room.

When I return, Fran requests polkas, country western, and religious music. We start with a gentle rendition of "Jesus Loves Me." It's just a warm-up. I dedicate the new song I've learned to Stanley and sing "One Moment in Time".

Our party is happening in the middle of a busy institution. A nurse charges in during the song to perform her duties. We keep going and include her in the party. Stanley sings along with me in a strong voice, using sounds instead of words. The chorus about "eternity" is strikingly appropriate to the moment.

The nurse rattles off medical jargon about Stanley's condition, but it feels disconnected, irrelevant at this point. We use her visit to take the opportunity to discuss Thyra and all the other great aides and nurses that have known and loved Stanley at Manteno. This nurse is dressed from head to toe in wild cowgirl regalia. She is real country and western and a member of the Boot Kickers Dance Club. We thank her for her help.

Tom: *This is one of Stasche's favorite country western songs by Ian Tyson, called "Navajo Rug."*

Stanley coughs and sings along.

Fran interrupts the song: *He loves parties. He responded to the song. He is just closing his eyes with the music. Oh, there he opened his eyes as if to say, "Where is it, where is it, where's the music?"*

Another nurse pops in to check on Stanley and encourages us: *What you are doing is better that anything we can do for him now.*

Stanley starts singing his own song and I say: *Okay, I'll sing the last verse.* But Stanley continues to sing his song. I mirror his singing with my vocalizations and extend his phrases. He is using a wide vocal range.

Tom: *I wonder if he is singing in Bohemian?*

Fran: *Helen (Stanley's oldest sister) tried Bohemian last week and he didn't respond at all.*

I quip: *That's because he wasn't singing. Stanley is the one who wants to sing!*

Fran suggests: *How about "Happy Days Are Here Again." Fran sings a rendition:*

> *Happy days are here again; the skies above are clear again;*
>
> *we can all afford a Henry Ford; happy days are here again.*

Tom: *Oh, I got a good one, "Lonestar."* Fran likes this one and I follow it with "Seven Spanish Angels". Certain phrases from each song catch in my throat.

Stanley starts singing with sounds again.

Fran suggests: *"Hinky dinky parlay voo,"* from "Mademoiselle from Armentières." We give it a try. Then I sing "The Marvelous Toy."

Fran: *Do you know "Hallelujah I'm a Bum?"*

> *Hallelujah I'm a bum, hallelujah bum again, hallelujah give us a handout to revive us again. I don't like work, work don't like me, that is the reason I'm so hungry. Hallelujah I'm a bum, hallelujah bum again, hallelujah give us a handout to revive us again.*

It goes on and on. It's the bum song. It's so familiar because the tune is from a hymn. There was another song, "He Had to Go and Meet her at the Astor." That's as risqué as it got in the 1930's."

Fran, going for something lighter and humorous, suggests: *How about 'Dirty Lill Dirty Lill lived on top of garbage hill, never washed and never will?" There is also "Dirty Will." Susan* (granddaughter) *knows the words to that one. The kids used to sing that Tarzan song. At the end of the song it goes: "Jane lost her underwear, me no care, me no care, Tarzan likes her better bare." These were two year old kids singing to Stanley who said, "What are you singing to me?!"*

Fran begins making moves to leave: *Stanley, we had a good visit didn't we? We sure had a good visit. I love you.* But we decide on two more songs: "Kiss the Girl" and the traditional calypso tune "Jamaica Farewell" which ends:

My heart is down, my head is turning around,

I've got to leave a little girl in Kingston town.

It's okay

Fran asks me: *Do you know "The Old Rugged Cross?"*

Stanley gives strong vocal and movement feedback.

Tom: *We need some hymns!* Fran and I struggle through "Holy Holy Holy." That is, we struggle with the verses, but we nail the chorus. This hymn is an old friend to individuals brought up in the Western Christian tradition in Stanley's generation:

Holy, Holy, Holy, Lord God Almighty
Early in the morning my song shall rise to thee

Stanley again sings along with even stronger vocalizations, and this time he raises his entire upper body off the bed! This is a man with end-stage Alzheimer's who has not sat up in over a year! There appears to be a deep spiritual connection between his process and the spirituality expressed by the hymn. He also seems to understand the words to the hymn, and may be forming the words in his mind despite his inability to physically express them clearly.

Fran says: *"Rock of Ages" . . . I don't sing it, only when I have to, but I remember it clearly because it was what they sang at my mother's funeral when I was six years old. Most people pick that one. But for him I am going to pick "When We Needed a Neighbor You Were there."*

Tom: *Well, tell him what you are going to do at his funeral.*

Fran: *That's why I like it because it fits our situation. When anybody needed a neighbor you were there. You were always there, Stanley.*

Tom: *Well, we should sing some more hymns. ~ Did Stanley ever mention what he wants at his funeral?*

Fran: ~ *He doesn't sing all those old hymns.*

I softly suggest: *Why don't you climb in bed with him?*

Fran's body jumps at the suggestion, but then freezes: *Stanley, can I climb in bed with you? . . . We can sing the "Doxology."* We stumble through it but pride ourselves on our effort.

Fran kids: *We don't want to sing "Praise the Lord and Pass the Ammunition", do we? How about "Beer Barrel Polka?"* The closest I can come is "Down at the Twist and Shout." *That was a pretty song Tom sang to you. He is a good singer. I bet you thought you were in Heaven already. It's okay if you go on ahead of me."*

Stanley visibly relaxes. The tension flows out of his body and his face softens.

Fran: *He looks so peaceful!*

Stanley as husband, friend, lover, provider, and protector has apparently been waiting for Fran to give him permission to leave, or for her to be ready for him to leave. He needs to know she will be okay. Why is this more than just conjecture? Because of his physical reaction in the moment, and because of Fran's confirmation when I offered to take her on this visit, on the seventh day of the death vigil. She said: *I know he is waiting for me.* In other words, Stanley could have died eight months ago when he was in a comatose state from pneumonia. Or he could have died seven days ago when he was in a second comatose state from stroke and pneumonia. Or he could have died yesterday or the day before. This begs the question, "What is his unfinished business?" This is a salient, deeply touching moment in the surprise party, the profound completion of a significant piece of unfinished business between them, a spiritual healing.

I exclaim: *Already there!? . . . Sorry, we can only walk you up to the Pearly Gates. We gotta check you in with St. Peter. We can probably only wave to Jesus . . . personal savior . . . I wonder who is over there that you are looking forward to seeing?*

This attempt to make a connection to the other side using Stanley's cultural background and spiritual beliefs create a long, thoughtful pause after which we revert back to a sing-along of "This Land Is Your Land."

Bed again

Stanley, my buddy, if you need to go, it is okay with me. If you want to stick around, that's okay, too. But I'll miss you if you take off. I have a premonition that you're leaving today. Last night in my dream,

Stasche told me that Stanley had died. *I think we had a good party. I love you, you stubborn old guy. Your son Stasche loves you and he says goodbye, too. If you have to go, he understands. And if you want to stick around he understands that too . . . Stanley, we gotta go. We gotta go. We will stay around for another ten minutes. If there is anything you need to say, or anything you want us to know, or anything you need to do yourself then do it now, because we have to go in ten minutes. But we'll be here for another ten minutes. Okay?"* I reiterate the ten minute deadline to make sure Stanley has every possible awareness about completing what he needs to complete.

Fran freezes: This is the last ten minutes with her friend, husband and lover on this planet. Then she says: *Tom's pretty good! You were with your grandmother before she died, weren't you, Tom?*

We sing one more hymn, "How Great Thou Art" by Stuart Hine:

O Lord my God! When I in awesome wonder
Consider all the worlds thy hands have made,
I see the stars, I hear the rolling thunder,
Thy power throughout the universe displayed.

Then sings my soul, my Savior God to thee,
How great thou art, how great thou art!
Then sings my soul, my Savior God to thee,
How great thou art, how great thou art!

Stanley, there is water in your eyes. This is evidence that Stanley is experiencing deep strong emotions again, in relation to his spirituality. There is no other physical cause in the moment, such as dust in the air, to account for spontaneous water in both his eyes. I do not try to interpret what these strong emotions might be. This simple observation supports him to feel what he is feeling.

I tell Fran: *I'll put my stuff in the van and give you some private time. Okay? Get into bed if you want to.*

Fran climbs into bed: *Look where I am, Stanley. I'm in your bed!*

I cheerlead: *Now you're talkin'!*

Fran: *I'm in your bed, Stanley. Did you know that? It's your girlfriend. I'm your girlfriend Frances, your sweetie. I'm in your bed and I'm going to hug you.*

Tom: *You can tell everybody you went to bed with him.*

Fran: *It's not as comfortable as I remember beds. They didn't have all this equipment on them. Can I put my head on your pillow? ~ He's got his eyes open a little bit.*

Tom: *Don't worry about the eyes. Do what you need to do . . . Stanley, I've got to go now. I'll see you on the other side. Hope you have a wild ride. Party on, Dude!*

I put my stuff in the van and lay down on the porch bench in the beautiful afternoon sunshine. Fran finds me asleep and says she left Stanley asleep too.

Tom: *That was fun. Stanley is the best audience I ever had!*

Fran: *What about me? I was listening!*

Tom: *Yes, but Stanley was singing along!*

As we climb into the van: *Well, Fran, the last time I saw you, you were in bed with Stanley. How was it?*

Fran laughs: *I didn't get much time because the doctor and the nurse came in. It was just like when the kids were home. Thank you, Tom for giving me this opportunity to say goodbye to Stanley.*

You know, when I get home and tell this story, nobody is going to believe me!

Twenty-four hours later, Stanley stops breathing in his sleep.

* * *

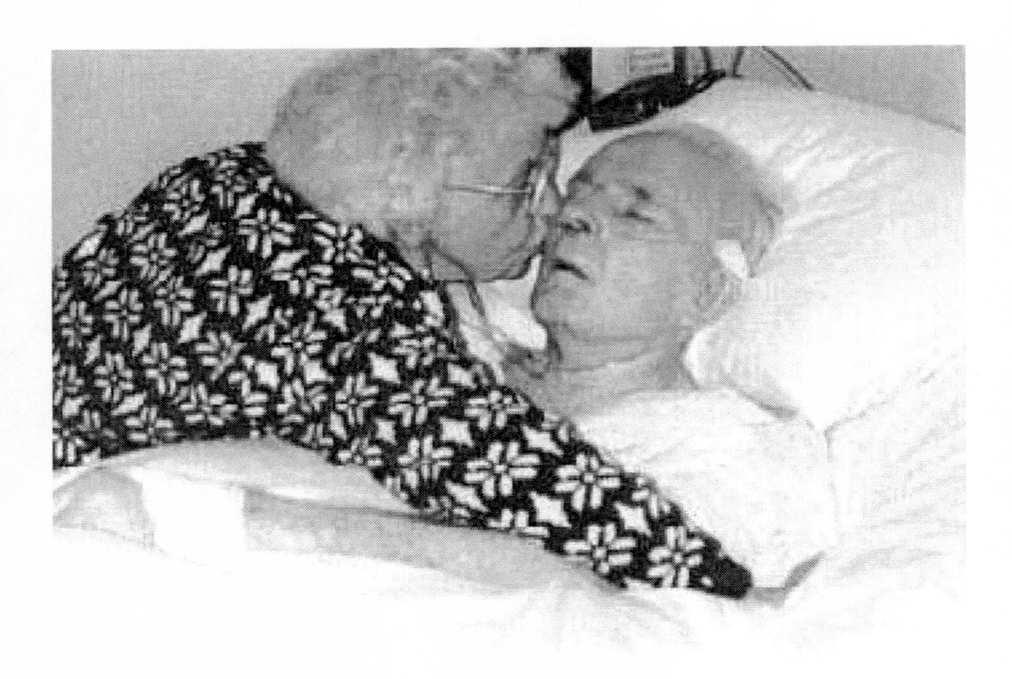

A goodbye kiss as Stanley sleeps at the end of his surprise party.

WHEN I NEEDED A NEIGHBOR

When I needed a neighbor, you were there

I was hungry and thirsty, you were there

I was cold and naked, you were there

I needed shelter, you were there

Wherever I travel, you'll be there

And the creed and the color and the name won't matter

You'll be there

Adapted from Sydney Carter

Chapter summary

1. **A boisterous surprise party** under the circumstances of a death vigil in a Midwestern, USA public institutional setting is a surprising idea in itself. Equally surprising is Stanley's participation. Indeed it is only possible because of Stanley's very responsive positive feedback. Initially he gives small positive feedback signals by opening his eyes a little wider and slightly nodding his head; these amplify into eye contact, vocalizations and relationship; and then into louder melodious vocalizations, significant leg movement, and large upper body movement, none of which have been witnessed for a year. The real "surprise" is the extraordinary extent to which he is present and participating and related throughout the party, despite his end-stage Alzheimer's state.
2. **Observing, commenting on, amplifying, and following** Stanley's communication signals in visual, auditory, body sensation, movement, and relationship channels, I follow his process, providing feedback that he is unable to give to himself. I thus facilitate both inner communication between parts of himself and his outer communication with us. His presence and participation arise from the disciplined application of these feedback techniques, rooted in a positive, supportive, loving attitude.
3. **Words:** Although Stanley cannot formulate words clearly, the words may be clear in his mind, similar to a stroke victim who knows what they want to say but can't get the words out. We can also guess into his vocalizations, and he can confirm our guesses with his positive feedback, like when he wanted more hymns. This is one form of binary (yes/no) communication.

4. **Music** is memorized by multiple parts of the brain and body which makes it a relatively dependable long term memory. For example, piano players can often reaccess forgotten songs by letting their fingers play them. As a musician it is likely that Stanley is actually remembering both melodies and lyrics to many of the songs we sing at his surprise party. For millennia music has been used in the healing arts of many cultures and in the *ars moriendi* (art of dying) practiced in hospices during the Middle Ages in Europe.
5. **Deep inner journey and relationship:** The surprise party is the last of many occasions when Stanley connects with his deep inner life journey and relates with those around him during his advanced Alzheimer's state. On every occasion during Stanley's advanced Alzheimer's state, if we started our visit by relating to where Stanley was, using sensory grounded sentient communication techniques, as taught by Drs. Amy and Arnold Mindell, and followed his deepest essential processes, he would relate to himself and to us and we would find mystery and intimacy and life and love.
6. **By sentient** we mean being subtly, sensitively, and finely attuned to perceiving minute flickerings of body feelings, movements, sounds, and images that catch your attention. One handy way to access sentient perceptions and sentient communication is to ask yourself, *What am I not noticing right now or almost noticing or just barely noticing?* Sentient perceptions and sentient communications are at the edge of our awareness. It is helpful to welcome them by holding yourself open to thoughts and feelings that are irrational and unexplainable, particularly first impressions.

7. **Sentient awareness** explores the messages, the dreaming, and the meaning contained in sentient perceptions and sentient communications.
8. **Finishing unfinished business** in Stanley's life on this planet includes assurance from Fran that she will be okay without his physical presence, and her permission for him to stay or leave. This is another "surprise" under conventional ideas about Alzheimer's: the advanced patient is considered incapable of this type of cognitive and emotional processing. People in all sorts of states of altered consciousness, given the awareness and opportunity, often check in with outer reality shortly before they die. The basic philosophy is that people require information from both inner and outer awareness before making major life and death decisions. A piece of Fran's unfinished business is to "let go" of Stanley and assure him that she will be okay if he decides to leave.
9. **Love:** The essential underlying and overarching content of the surprise party, of course, is love between Francis and Stanley, supported by the love of Tom and caregivers at the Veterans Home. The love present in the room is enhanced and communicated by noticing and supporting small nonverbal and verbal signals. This is the love story of one person's leave-taking. Every family will have their own style of expressing love near death.
10. **Creative sacred ceremony:** The surprise party becomes a creative sacred ceremony that offers celebration, blessing, honor, and closure for Fran and Stanley, their family, and friends. This points to the potential for more families suffering from Alzheimer's dementia and other end of life altered states, to experience this kind of deep ceremonial closure and spiritual healing: facilitated, yet spontaneous.

Chapter 1 Exercise: Creative pre death celebration

Created by Stan (Stasche) Tomandl

Work to your own comfort level in the following experiential exercise. Thanks for taking the courage to engage in this essential work of the human spirit.

1. Imagine you are on your deathbed. Briefly write down what elements you would like at a pre death celebration: people, music, food, ceremonies, etc.
2. Next imagine that you are dying, by lying down, relaxing, and taking a few deep breaths. See yourself, listen to yourself, move a bit like you think you would move, breathe how you would breathe, and feel into your near death body.
3. Remaining in that state, pause a moment and praise yourself for the important accomplishments of your life.
4. And then say, "I am going to die having done everything except ____________________________."
5. Imagine what help you would need to accomplish this. Make a note.
6. Check to see if the requirements for this accomplishment are included in your pre death ceremony in step #1. If not, add these ingredients to your celebration plan.
7. How might you incorporate these celebration elements into your daily life? Your work, relationships, recreation, etc.
8. How would you view the world differently? How would the world view you differently?
9. Thank-you for participating in making our world a more livable place.
10. Remember to review and renew your pre death ceremony once a month or however often you need to.

Exercise notes

Chapter two

Nobody Is Going to Believe Me

On our way home from the surprise party Fran says: *You know when I get home and tell this story, nobody is going to believe me!* Fran and Stan experienced a profoundly intimate spiritual healing. And yet why is it likely that "nobody" is going to believe her? It could be that what she experienced is considered impossible by conventional Western medical standards.

According to the Alzheimer's Association of the United States, "Alzheimer's disease, the leading cause of dementia, is a progressive brain disorder that gradually destroys a person's memory and ability to learn, reason, make judgments, communicate, and carry out daily activities." If this is the case, then there would be little possibility for Fran to communicate or relate to Stanley in his most extreme end-stage Alzheimer's state, much less party with him, sing songs, share spiritual experiences, and complete their unfinished business. And yet they did. Stanley's "Alzheimer's Surprise Party" is an illustration of five years of applied curiosity, love, and close observation while following and supporting him in his remote states of altered consciousness.

Research hypotheses

After twenty years of working with individuals in early to advanced stage Alzheimer's dementia and coma, using a progressive form of communication and awareness facilitation. we believe Alzheimer's and other dementias may involve a complex soulful and spiritual growth process. This process lives in the background of what otherwise appears to the conventional observer as unrelenting deterioration into a "mindless pathology." We would like to propose:

1. Even in the most advanced dementia states somebody is home, people are not altogether gone.
2. They are trying to communicate with themselves and the outside world to the best of their ability, often with hitherto uncommunicated messages.
3. Two-way communication is possible even with people in the most advanced dementia states and coma.
4. Relationship is also possible even in the most advanced dementia states, including delirium and coma.
5. There exists a complex soulful and spiritual growth process in the background of people with Alzheimer's dementia that can be categorized into the following processes: a) working on "unfinished business", such as resolving individual and family issues, including meeting new aspects and partially lived aspects of the personality; b) "harvesting", such as recalling and savoring life experiences; c) "imparting blessings", such as openly accepting loved ones; d) creating "sacred space" for marginalized experiences by tapping into a sense of something larger than ourselves; e) garnering "meaning" by exploring formative experiences and essential beliefs; f) making "spiritual connections" like immersing in the beauty of eternity; g) intimacy and deepening of relationships; h) working creatively on cultural and societal conflicts. These processes are not linear stages of Alzheimer's dementia, but rather nonlinear descriptions of currents that may run separately or concurrently.
6. Caregivers can help facilitate inner awareness of the above processes in people experiencing early to advanced stage

Alzheimer's dementia for their benefit and the benefit of the family and society.

7. Detachment from personal history in people with Alzheimer's dementia is normal, but not necessarily permanent. Personal history may remain intact although access and motivation to access it are a function of patients' growth, interests, and course of the disease process.
8. While personality and behavior can change dramatically as people in Alzheimer's dementia states evolve; inherent character traits, beliefs, and memories can remain intact both pre and post Alzheimer's dementia and offer portals for entering into communication.

Process Work and awareness

Our progressive form of communication and awareness facilitation comes from Process Work, also known as Process Oriented Psychology, based on the pioneering work of psychologists Drs. Arnold and Amy Mindell and their associates. At the heart of Process Work is experiential training in sentient awareness facilitation.

If deep sentient awareness could be bottled, it would be hailed as a "miracle drug" for the body, mind, and spirit. The more a person experiencing Alzheimer's dementia, and his or her immediate family and community, can be aware of the deep underlying meaning and direction of dementia communication processes; the more inner resources and outer interventions become available to all people involved for the benefit of patients, family, caregivers, and community. Sentient awareness does not replace medical care. Sentient awareness is an act of reverence that complements and enhances the whole person; mind, body, emotion, environment, and spirit. In our personal experience, Process Work communication and sentient awareness facilitation is effective for communicating with and relating to those in Alzheimer's dementia states.

Background and benefits

Dementia is a loss of mental function in two or more areas, such as language, memory, visual and spatial abilities, or judgment, severe enough to interfere with daily life. Dementia itself is not a disease, but a broader set of symptoms that accompanies certain diseases or physical conditions. Well-known diseases that cause dementia include Alzheimer's disease, multi-infarct (or vascular) dementia, Parkinson's disease, Huntington's disease, Creutzfeldt-Jacob disease, Pick's disease, AIDS, alcohol and drug abuse, and Lewy body dementia. Alzheimer's accounts for approximately 60% of people with dementia, vascular dementia accounts for approximately 30%, and all the rest account for approximately 10%.

In 2005 the first North American baby boomers turned 60. In only twenty-five short years, every other baby boomer, statistically speaking, who reaches the age of eighty-five will have some form of dementia. In total, it is estimated that 10 million baby boomers in the US will develop Alzheimer's.

Currently half of all nursing home residents have Alzheimer's dementia. More than 7 out of 10 people with Alzheimer's dementia live at home, where almost 75% of their care is provided by family and friends.

Worldwide there are currently 26.6 million identified Alzheimer's dementia patients. If we make a few basic assumptions and add 26 million significant others, 52 million children of Alzheimer's patients and 26 million siblings and an equal number of 104 million people servicing Alzheimer's dementia patients, the total comes to 235 million people worldwide affected by Alzheimer's dementia.

Alzheimer's dementia is a pandemic. It is everywhere, it is big, and growing bigger.

Benefits of Applying Process Work

For people in early stage Alzheimer's, the benefits of our research work, as expressed by grateful clients are:

- The feeling of being aided and understood, instead of remaining psychologically and emotionally isolated and helpless;
- The opportunity to go farther into knowing and expressing themselves on their own terms in their new states of consciousness.

For family members, staff members, volunteers, and administrators, the benefits can include:

- Reduced feelings of hopelessness and burnout common to Alzheimer's dementia caregivers;
- Less fear, added peace of mind, and reduced stress from greater awareness of larger spiritual processes;
- The opportunity of being in closer relationship with affected loved ones; and
- The opportunity to establish alternative communication patterns in early stages, which can be useful during the more advanced stages of Alzheimer's dementia.

The benefits of Process Work for the health care system may include:

- Techniques for relating better with people in early stages;
- Techniques for managing patients' agitated and extreme behavior through more effective communication, with less need for force or physical or pharmacological restraints in later stages; and
- Improved two-way communication regarding care of patients in remote and comatose states.

For society, the long-term benefits of Process Work may be both direct and indirect:

- Direct benefits could include more compassionate and effective forms of patient management and care with reduced expense for the long-term care of Alzheimer's dementia patients;
- Indirect benefits may include reduced stress on employees, resulting in lower cost of employee benefits and less downtime and stress leave.

The immense tragedy and huge expense of caring for those in Alzheimer's dementia states of consciousness have made Alzheimer's dementia a major social and ethical issue.

Because of our experience, we want to raise awareness and help relieve the suffering and stigma for this large universe of people with Alzheimer's and other dementias, and for their families and caregivers. We want to share many ways of connecting deeply through entering their worlds. We want to tell the story of Stanley Tomandl in depth, as an example that we hope inspires more people to open to the gifts and blessings that go along with this intense and difficult time in people's lives.

Chapter Three

My Father My Journey

Mrs. Nischke loved her rose bushes. One evening during one of my parents' many neighborhood dinner parties, she casually mentioned that she needed a few old Christmas trees to cover her rose bushes to protect them from the winter frost. That was all Dad needed to inspire the trickster in him. By morning he and my mother, his ever-willing accomplice, cooked up a scheme to "help" the Nischkes. Dad placed an advertisement in the local paper that read:

Used Christmas trees wanted

for protecting rose bushes,

drop on front lawn at 2223

Walters Ave, Northbrook

The ad drew a tremendous response. By the next morning Nischke's front yard was piled high in a mountain of dead Christmas trees. A frantic Mr. Nischke was told by Police Chief Glenn Ford: *We can't arrest Stan; hell, we put our trees on top of the pile with all the others.* And his complaint to the newspaper office also fell on deaf ears: *It's the funniest thing that ever happened around our newspaper office!*

Of course Dad, sly fox that he was, didn't undertake his Christmas tree caper without dropping a hint to the police chief beforehand. And he also arranged for cleanup. When my parents and their friends picked up the last Christmas tree to load it on the clean up truck, they discovered a bottle of whiskey with a note taped to it: *Come inside for a drink!*

This escapade was typical of my father, Stanley Tomandl, always a spontaneous character. Against a background of good humor and fun

Stanley lived many lives: farmer; carpenter; husband and father; undertaker; ambulance driver, soldier; musician; community leader and activist; interior designer; and spiritual seeker. He deeply enjoyed many pastimes, including: fishing; hunting; gardening; card games; dancing; parties; and practical jokes.

But by age 71, eleven years after a cerebral hemorrhage, it was clear that those days were over. Dad's memory was not good and, in my mother's words: *The happy times are slipping away.*

My father's journey into remote sates of consciousness began with a massive cerebral hemorrhage at age 59. Over the next ten years he recovered normal consciousness to about 90% of his former capacity. Dad worked and played and traveled, continuing to connect with family and friends. At age 69, he began to go back into progressively remote states, starting with confusion and memory loss, then Alzheimer's dementia, leading to semicomatose states with occasional metabolic coma.

I remember my father upon entering the nursing home at age 72, three and a half years before he died. He longingly said: *I wish things were the way they used to be.* As his son I fearfully wondered: *Where has my father gone? That's not him! I want him like he used to be. He was so vital, so full of life. He's not here anymore. He doesn't even remember me, his own son! I need him. I want him back!* Tragedy, hurt, grief, shock, anger, bewilderment, and why?

These reactions are one part of my experience of a much larger and deeper story. I could have just dwelled on my own misery and grief and railed against Alzheimer's, but I also chose to follow and honor Dad's path into his remote states of altered consciousness. By doing so I shared parts of his journey with him and explored amazing landscapes of love, relationship, beauty, and profound spiritual meaning.

I discovered that Alzheimer's dementia is way more than a pathological condition; it is something much larger. In the words of a woman who lost her father to Alzheimer's and has had much time to contemplate her tragedy: *Alzheimer's is going to teach us a lot!*

I believe one of the reasons many of us get stripped of our recollections is to allow other experiences to emerge and take form, perhaps even beautiful experiences. And while we may look with pity upon advanced Alzheimer's patients, with close observation of my father over time, I seldom detected pain signals; only temporary hurt expressions, no groaning or crying out. He lived in a different state of being, a parallel universe to our daily lives in consensus reality. When we helped facilitate his awareness of both his more normal states of consciousness and his farther away states, he became more communicative, more related, happier, and more content.

Families and friends can suffer immensely with a loved one in Alzheimer's dementia. My own and others' trauma and incomprehension about who and what our loved ones become, is nearly impossible to bear at times. But what is it like to be on the inside of the Alzheimer's dementia journey? Who is in there? How are they different to themselves? Where are they going? How do we communicate with them? *How do we help them communicate with themselves?*

Over the past twenty years my longtime friend and colleague, Tom Richards, myself, and my wife Ann have researched process techniques to explore some of the mysteries of Alzheimer's dementia. Our process approach connects people in Alzheimer's states with the beauty behind their outward Alzheimer's experiences, and looks for meaning for the benefit of patients, families, caregivers, communities, and society.

As radical as it sounds to equate beauty with Alzheimer's states, we discovered beauty both in our direct experience in relationship to my father, and in his own inner process, even in his most extreme emotional or extreme remote Alzheimer's states. And we were startled to hear the word "beautiful" in his clearly spoken description of his experience near the edge of death.

This is a love story. It is the story of the anguish and ecstasy of trying and succeeding, beyond all conventional wisdom, to stay in communication and in relationship with a loved one who is in and out of very remote states of consciousness on their spiritual path. It is a story of human hearts searching, honoring, surrendering, celebrating, and embracing a process larger than ourselves.

Chapter Four

Fifty-First Wedding Anniversary

Stanley could pull outrageous pranks, and people would usually love him for it. He was a natural comedian with an excellent sense of timing. He often used humor as a tool for intervening in socially explosive situations. He could bring out "unmentionable" topics in groups, relieving tensions and improving moods. Other times he just liked to cut up. There are a lot of great funny stories about Stanley.

Fran: *One time we were riding with some friends. They drove way too fast, so Stanley grabbed the St. Christopher (patron saint of safe travel) medal off the rearview mirror and threw it out the window exclaiming, "The way you're driving it's not going to do us any good!"*

Fast-forward

In this first published volume of our experiences and research findings regarding Alzheimer's dementia, we also move fast. Please indulge us with good humor and open hearts and minds as we now fast rewind to two years before the end of Stanley's life on earth, and recount stories from this period in great detail. The early stage and intermediate stage of his Alzheimer's journey will be presented in subsequently published volumes. In our present chapter we step into our story on Fran and Stanley's fifty-first wedding anniversary. Stanley has advanced Alzheimer's dementia and has been living in the Veteran's Home in Manteno, Illinois for two years. Their fifty first wedding anniversary takes place more than two years before his surprise party and death.

Anticipation

It is November 23, 1992, a few days prior to Fran and Stanley's fifty-first wedding anniversary. With fear and trepidation I offer to take Fran to celebrate their anniversary. I am not optimistic about our visit because of my lack of confidence about facilitating their relationship interactions, using the multilevel, multichannel awareness techniques of Process Work that I have been studying.

Fran is also not optimistic; it is not at all the kind of anniversary celebration she had dreamed about for their retirement years, and because of her experience last year on their 50th anniversary, which they also celebrated in the Manteno Veterans Home.

Stanley Tomandl, American Legion Post 791 Past Commander & Fran on their 50th anniversary, Manteno, Illinois 1991~ drawing by Ann Mah

Fran: *On our 50th wedding anniversary I was pleased and appreciative when Thyra, the nurse's aide who can handle Stanley the best, came in to dress him on her day off. With another aide's help they put a suit and tie on him. He always liked to dress sharp. We had all three of our kids, Jeanne, Stasche, and Dan there, and Ralph (Jeanne's husband). We ate ice cream and cake. It sure wasn't like the 50th I had imagined a few years ago!*

Coaching

On the car ride down I review my notes from previous visits with Fran. She wants to know the game plan. I tell her we will go slow, pay attention to Stanley's signals, mirror them back to him, follow his feedback, and watch for cycling, which would mean he wants to go farther with the issue he keeps repeating. I warn her that two people conversing in front of Stanley will confuse him.

On arrival we locate Thyra, Stanley's primary caregiver, who reports Stanley had undone a button on her blouse and learned a new trick of putting his head on her shoulder, while putting his hands on her hips. For many people in Alzheimer's dementia states, human contact, including sexual contact, is one of the last things to remain in their consciousness. This can be difficult for many family members and sometimes for staff. Fran is very matter-of-fact about it all. Thyra also reports that the head physician and Stanley have been talking in Bohemian (Stanley's first language).

The "attack"

We find Stanley in the TV room in his geriatric chair with his head down, eyes closed fast asleep. Fran immediately rubs his shoulders, says hello, tries to jog him awake, kisses him on the lips, and demands that he open his eyes. I am standing there stunned, shaking my head. Because Fran is so obviously overwhelmed with pleasure and excitement at seeing Stanley, our so-called preparation has just gone out the window. To my surprise, in spite of the suddenness of this sneak attack, Stanley kisses her back, but refuses to open his eyes on Fran's command. He takes what he likes and ignores the rest. He belches and starts his recurring chewing motions. I wheel him in between the TV room and the nurses' station where we can have some private space, except for the residents' public pay phone nearby. Fran puts Stanley's glasses on and he opens his eyes.

Making love

Tom: *Richards here now and Fran is with me. Is it okay to visit?* Fran has grabbed both his hands and I explain that to him: *Fran is holding your hands.* I ask Fran: *Are they cold?*

To my surprise Stanley answers: *No, I don't think so.*

They hold hands for less than a minute when Fran suddenly lets go and pulls her hands away. Stanley is breathing very deeply with his head down, indicating a feeling state. Then he lifts his head and says: *It doesn't.*

Tom to Stanley: *Do you want to hold hands?*

Stanley: *Yeah.*

Tom to Fran: *Why did you stop holding hands?*

Fran: *Hurt my hand.*

Tom to Stanley: *Be gentle with your wife.* Fran has sensitive arthritic hands. I suggest to Fran that she move closer to Stanley and get more comfortable. Fran moves her chair side by side with Stanley's, like a love seat arrangement. They have always been a passionate couple.

They are focused on each other and ignore me. They hold hands quietly for several minutes. Stanley seems to simply enjoy the ecstasy of this quiet intimacy. Fran, however, is experiencing strong reactions. Her face is growing flushed and her posture is mixed; part wants to stay and enjoy the intimacy and another part seems scared, ready to pull away again. Her head and hands want to stay, but her shoulders are poised, ready to pull back.

Stanley smiles as he calmly says: *Dates real good!* Then he speaks in Bohemian. The switch to Stanley's birth language likely indicates a difficulty or shyness to be more direct in relationship to Fran.

Tom to Fran: *You are intimate and he loves it!* Stanley smiles.

Fran: *I don't want to cheat on Thyra.* I am flabbergasted! This is such an amazing statement that even today, writing this, I hardly know what to say about it. Isn't love fascinating! Thyra could represent a pattern in Fran's psychology to help her get over shyness about greater intimacy with Stanley in an institutional setting. Thyra would be a better "partner" for Stanley because she knows how to relate in institutions.

The pay phone rings and Fran asks: *Should we get it?* She doesn't want to let go of Stanley's hands in the moment, so I get up to answer it. Fran and Stanley are now doing a lot of kissing and as I go for the phone I cheer them on: *Keep making love!* Romance and physical intimacy is not only the domain of the young. Elders experience the same stirrings, and our culture does not support this basic and sacred human need enough in institutions.

Fran backs off first by changing the subject. She asks Stanley: *Did you get a bath?* Stanley was leading Fran to greater intimacy. Her pulling back is perfectly natural and serves a purpose. Each time

Fran tests her freedom of choice to avoid, she gains more confidence and a sense of safety to go farther.

Stanley: *I don't know.*

Fran now realizes that she has backed away from a good thing and wants more. She kisses him a few more times from her chair and says to me: *He usually pecks me back, but NOT LIKE THIS! You can feel the pulse when you hold hands like this.* She gets up out of her chair for another kiss. Stanly parts his lips and they exchange an extremely romantic, long, passionate kiss! Fran is blushing scarlet and I have to glance away.

Stanley sums up the experience and says to me: *One nice thing, it's real good the way she's coming along!* I burst out laughing inside. I take it as Stanley, the identified "sick" patient, evaluating and diagnosing Fran, the "well" person. In the moment sickness and wellness and labels like "Alzheimer's" have no meaning. <u>This is life at its best, intimacy and ecstasy!</u> Stanly adds: *She was.*

Fran concurs with him: *Yeah, she was.* Stanley looks over at Fran. I am amazed at his ability to relate to both of us in the moment.

Tom: *Today we meet The Lover!*

Fran wants more: *You want to kiss me instead of chewing! Are you wetting your lips so you can kiss me?* She gets out of her chair for another kiss. It is a big one, but then they both withdraw. I am reminded of something that amazed me the first time I heard it, that approaching ecstasy can be as scary, if not more scary, than approaching the dark side of human nature. So we spend most of our life "stuck", slumming around somewhere in between the dark side and ecstasy, with only glimpses of what could be. I feel privileged to share this exceptional moment of intimacy with Fran and Stanley.

Making war

Fran asks Stanley: *Should I tell you a joke?*

Stanley: *Yeah!* He smiles.

Fran: *There's this couple that is expecting a baby. The doctor instructs the Mrs. to bring back a specimen on her next visit. She goes home and asks her husband, "What is a specimen?" He says, "Piss in a bottle." She says, "Well shit in your hat!" and the fight was on again!* In the background is the fight. Misunderstandings and power struggles form a natural balance to attraction. Today, however, we enjoy the love and intimacy that is deep and whole.

Both dead

We all laugh and Fran sings a favorite old ditty:

> *O'Reilly is dead and his brother don't know it; His brother is dead and O'Reilly don't know it; They're both of 'em dead and lying in bed, and neither one knows that the other one's dead.*

Fran sings several refrains and then explains to Stanley: *Their problem is that they're both dead.*

I imagine this as more than a song. It seems a normal reaction to what has transpired. Having just experienced ecstatic moments, the rest of life must feel dead and "I didn't know it," just like O'Reilly and his brother.

Tenderness

Stanley and Fran gaze into each other's eyes and hold hands nonstop for more than three minutes before Fran becomes uncomfortable and lets go of Stanley's hands.

Tom: *Fran, why did you let go of his hands?*

Fran: *So you can talk to him. I should share him.* Stanley's reaction to this is to completely shut down and go remote, head down, eyes closed, deep chest breathing. Then he starts rubbing his fingers. I suggest to Fran that she replace one of Stanley's hands with her hand and rub his first finger the same way she sees him doing it. I explain that by doing this, Fran will be taking over one part of a two-way communication that is taking place between Stanley's fingers.

As Fran is doing this she asks Stanley: *Does that mean something, kind of naughty?* Stanley rubs faster, indicating positive feedback. Fran is on the right track.

Tom: *How about REAL NAUGHTY!*

Stanley: Rubs his finger very fast. We all laugh. At that moment someone pushes a cart past us. Stanley looks away, watches it as it passes and lets go of Fran's hand. Fran rests her hand on his leg to see if he will take it again. He takes her hand and lifts it very carefully. He examines it. As he is gingerly rotating it, he begins a gentle massage. He seems to remember that Fran has arthritis in her hand and is now tenderly massaging it.

Get that lady

He does this for a long time and then gazes down. I respect the quiet intimacy until the moment he asks: *Do we need to look down there? Most of it will be gone.*

I repeat his words back to him: *Yeah, most of it will be gone.*

Stanley: *I don't know where it's going.*

Tom: *Are you talking about your mom?* Last time he looked down there during a past visit, he "saw" his mother. In retrospect, I am off the track. As communicators we sometimes try to divine the communication flow by relying on past experience, but then must take in moment to moment feedback to check out the validity of that direction. Based on his following statement, I believe "it" is a reference to himself back in his old home in Northbrook.

Stanley: *Yeah. Beautiful high desk all.*

Tom: *Big high desk to do work?*

Fran breaks in: *He hated paperwork at his desk.*

Tom: *Oh, paperwork on our book at your desk! I get it!* In previous visits we had discussed this book you are reading, and he recalled his big desk in Northbrook. I could be wrong, but I do not get a chance to explore it because Fran wants to read him his Thanksgiving card. However, she doesn't get a chance.

Stanley: *Get that lady.*

Tom: *Get that lady and love her?*

Stanley: *That's right! Drive it out there . . . all the stuff.*

Tom: *What to do with all that stuff? Throw all that stuff out?*

Stanley: ***Yeah!***

The hug

Tom: *Can I give you a hug?*

Stanley: *It's hard to know.*

Tom: *I'm going to give you a hug anyway.* I hug him.

Fran: *Boy, did he react to that! I usually hug him but he never hugs me back.*

Tom to Stanley: *Can you hug Fran?*

Stanley: *I think I can do that.*

Fran and Stanley hug and Fran asks: *You like that?*

Stanley: *Yeah!*

Tom: *Well, that's a real high note to end our visit on! You keep working when we're gone.* Fran and I say our fond goodbyes to Stanley and depart for Northbrook.

Teacher

On the way home I comment on how well they are working on their relationship. Fran is taken aback. It is news to her that they are still working on their relationship!

Fran: *You say you are learning from him. He is your teacher. I never thought of it that way.*

Alive

After I drop Fran off, I have to return; she left her change purse in my car.

Tom: *Fran, you left your change purse in my car.*

Fran: *And the garage keys, right?*

Tom: *Yeah.*

Fran: *That happens when you get old.*

Tom: *No, it happens after you've been stimulated, after you've been alive!*

Alzheimer's intimacy

On reflection, I think this was one of Fran and Stanley's most intimate moments in the Alzheimer's phase of their relationship. **Love keeps loving.**

Chapter Summary

1. **Two anniversaries:** The celebration of their 50th a year earlier was a cake and ice cream affair, with lots of family, but little passion. I suspect that their 51st was one of the most intimate anniversaries in their relationship, Alzheimer's or not. In fact, I have to revise that statement, because new styles of intimacy can be one of the gifts of Alzheimer's. It probably wouldn't have unfolded the same way without the Alzheimer's altered state to eliminate barriers, and outside facilitation to help them be more aware of their love and intimacy.
2. **Extraordinary things** occur during this visit. Stanley occupies his Tough Guy role by refusing to open his eyes, but he accepts a kiss. The kissing is very passionate, to Fran's astonishment and delight; Stanley is more open to intimacy than Fran. He says: *It's real good the way she's coming along!* Stanley relates to both of us without confusion despite my advance coaching with Fran to the contrary. Stanley quickly returns my hug.
3. **Relationship:** Fran is astonished to learn that they are still working on their relationship, which we also witness near Stanley's death during his surprise party.
4. **Love story:** It occurred to me while rereading this chapter, that if I were to remove my comments and write just the dialog as a screenplay, it would resemble a passionate Hollywood style love scene, and that readers would have no idea Stanley had dementia or that we were in the hallway of a veterans' home. Love happens where love lives.

Chapter 4 Exercise: Ecstasy and Alzheimer's dementia

Created by Stan (Stasche) Tomandl

Work to your own comfort level. Thanks for finding out more about Alzheimer's dementia and ecstasy for yourself and others.

1. Recall a time in your life when you were in an ecstatic experience or you almost had an ecstatic experience or you wanted an ecstatic experience, or failing the above, imagine into wanting one. The ecstasy you desire may possibly be from sports or intimacy, relationship or family, related to creativity or business, based in spiritual practice, or brought on by being in nature or with a presence that is larger than yourself.

2. Now find out more about ecstasy by taking it farther. To do this, imagine you have Alzheimer's or some other form of dementia. With all due respect, you may remember someone who has dementia and become like them, or you can imagine into your own version of someone with dementia.

3. Look at the one with dementia, listen to her or him, then sit, stand, or lie down like they would. Next feel into how they feel in their body, and then move a little bit as they would move.

4. Now make a quick line sketch of that movement.

5. While maintaining your Alzheimer's state, squint at your sketch through half closed, unfocussed fuzzy eyes. Now write down what the sketch looks like, the first impression that pops into your mind. There might be a story or short video that rolls out of that impression. Briefly write this down, too.

6. Go back to your ecstatic experience in step #1 and explore how the energy and pattern from your sketch, impression, and story can help you enter and exit ecstatic experience more thoroughly and satisfactorily.

7. How would this energy and pattern help you in your daily life of work and relationships?

Exercise notes

Chapter Five

Fantastic Person

Fran: *One night many years ago, we threw a party and during the party one of our friends, Don, who owns a tavern, says, "Well, you know, after all these years, we don't even have any draperies in our tavern." After Don leaves our party, Stanley goes down to the drapery workroom in our basement.* (Stanley lived many lives before and after his military service, one of the most successful being his interior design and drapery business with Fran).

He scrounges up all kinds of gaudy old fabric. So, it's like one-thirty in the morning, and Don is hosting a fish fry at his tavern. Emil and Stanley go down to the tavern. The tables are full. People are dining. Stanley walks up to a table and says, "Pardon us, we're putting up draperies!" Emil hands Stanley a staple gun and this gaudy fabric. Stanley climbs on the top of the table, dishes and all. Standing in the middle of their dinner, he staples this ugly fabric to the cypress wood paneling. Amazingly, Stanley made new life long friends with the folks at that table . . .

Introduction

This chapter reveals how Dad travels from minimal signals in one of his more remote states of altered consciousness to an intimate exchange with me. For two hours Tom and I intensely and patiently follow Dad's heightened sensitivity. Often minimal signals indicate a high degree of sensitivity coupled with a reluctance or inability to

Anti Aircraft Artillery Corporal Stanley Tomandl, 1945 El Paso, Texas

* * *

respond to outer stimulation. Some caregivers and loved ones may mistake minimal signals or no response as lowered sensitivity. There may be fear of the unknown, or a lack of patterns to communicate in new ways. Great patience is required to work with minimal signals from someone in a deep withdrawn state. Reading the following detailed cue by cue account may take patience. Yet underneath the slow rhythm of Dad's retreats and approaches, lies an exciting adventure of intimacy and connection. Please, follow us further.

Just shoot me

Tom and I drive out to Manteno in his wonderful old van with captain's chairs and oak trim. It's August 9th, 1993; Dad has lived at the Illinois Veterans Home for two and a half years now. I'm in Illinois for a couple weeks to visit my folks, especially Dad. Here are some of my myriad inner thoughts while on the drive through miles of cornfields: How much longer will he live? And in what state of mind? Will he relate with me? I want to know him better as an adult. This may sound bizarre given that he can't even remember my name to speak it. But we can interact using the minimal communication signals that he makes. In fact we have related in ways we never could pre Alzheimer's, thanks to personality changes in his Alzheimer's states of consciousness. I have brought up subjects and evoked emotions that previously have been uncomfortable for both of us, and will do so again today.

I have driven out to Manteno every other day. Visiting Dad is hard on me: the three hour round trip in hot weather with all my doubts and feelings. Could I possibly do more for him? I want to help Mom and him connect more, for her sake especially, and can't, despite all my skills. I'm sad. I'm angry. I'm exhausted and discouraged. Fortunately, I also feel full of love and enjoy hanging around with my father in his far out states.

On the other hand, I find myself thinking: *If I'm ever in this state, try to work with me for a short bit, and if it's no go, there's little communication, just take me out and shoot me.*

If we (I, Tom, colleagues, medicine, psychology, our culture) had more awareness and skill, could we work better and quicker with Dad? It is an agonizing question for me.

I also realize nature has its own pace, human nature included. My mom and dad, the extended family, the institution, Tom, and I are all part of nature and we all need time. We can only learn and change at a certain pace. I'm glad Tom is coming along today and that we're videoing. I need outside input, feedback, friendship, and support.

Hello goodbye

We arrive and set up the video equipment while a nurse administers pills and water to Dad. She attempts to communicate verbally, but Dad does not respond verbally. He does, however, get the message and swallows his medication followed by water.

Dad is lying in bed, his upper body elevated about thirty degrees, a pillow under his head. He appears stiff in the neck and shoulder area, holding himself slightly off the bed and pillow. He gazes at the far end of the ceiling. His expression is neutral. His hands lie in his lap under a light blanket. His legs are drawn up a bit sideways:

Stasche: *Hello, Dad.*

Stanley: *Mumble*, makes eye contact with me.

Stasche: *Ah ha.*

Stanley: *Mumble.*

Stasche: *Yeah, hi. It's your son, Stasche.*

Stanley: *Oh.*

Stasche: *Yeah. Hi! Hi!*

Stanley: *Mumble.*

Stasche: Yeah, it's me. How you doin'? Pause for response.

Stanley: *Mumble.*

Stasche: Laughter. *Yeah.*

Tom: *Richards here.*

Stanley: *Mumble.*

Tom: *Yeah, how the hell are you doing?*

Stanley: *Mumble.*

Tom: *Yeah.*

Stanley: *Mumble.* Dad has greeted us briefly with eye contact, the word "oh" and assorted "mumbles" that seem to represent attempts at words. This is a fairly typical initial contact with Dad. He pops out briefly to say "hi" and then goes back to his usual more remote state of altered consciousness.

Tom to Stanley: *Mm mmm . . . you're doing the chewing. ~ I would actually hold his hand.*

Stasche: ~ *Let's go back and work with him where he's at. ~ We're going to talk about you, Dad, as we're going along here.* Slight pause in Dad's chewing rhythm, possibly indicating a response to our comments.

Warming it up

Stasche: *And sometimes we'll be working very personally and getting the "shit" out.* By this I mean Tom's and my personal emotional reactions to Dad and the whole situation. I wish I would have explained this more explicitly at the time. Pre Alzheimer's dementia

Dad would have understood this meaning of "getting the shit out." But since he's in a different state, we shouldn't assume he knows, especially since he shows no reaction to this statement. So Dad's probably not interested in our shit, but Tom and I have to get it out anyway for our mental health and emotional well being.

Other times we're going to notice what's going on with you, Dad.

~ *I'd work with the chewing.* ~ Dad chews throughout most of our visit. This looks like a very methodical mastication of a good sized mouthful.

Tom: *Chewing is good.* ~ *I'd like to see his hands, though.* ~ In Tom's experience, Dad "works" with his hands a lot, touching and folding fabric and paper.

Stasche: *Oh, get some hands exposed, good idea, Tom. We're going to take this cover off. Here goes, see what your hands are doing under here. Oh, your hands are nice!* ~ *You pick up that hand; I'll pick up his other hand.*

Tom: *I'll pick up your right hand.*

Stasche: *Yeah, there, put the cover under there. There you are, Dad.*

Tom: ~ *While you're working with the chewing, I'll hold his hand.* ~

Stasche: *I'm going to put my hand inside your hand. Great! Nice to make contact.* Chewing stops momentarily. During this whole interaction of the hands, Tom and I communicate sensitively with Dad, trying to let him know what is going on. He exhibits no change in his behavior until I put my hand inside of his, then he stops chewing momentarily. Our hands connecting have given Dad pause. This is an early hint at what transpires later with his hands and me.

Tom: *Okay, I'm going to take your left hand, okay. That's Richards right there. How you doing, buddy?*

Goin' fishin'

Speaking in a slightly higher pitch, Tom now attempts to imagine into and give voice to Dad's inner state. We watch for any reaction, big or small, that would indicate which bait hooks Dad's awareness fish.

Tom: *I feel a little rigid.* Tom uses words that reflect Dad's rigid muscles. *I'm feeling a little apprehensive right now, not too sure what's going on.* Even though Tom and I have taken extreme care to move Dad's hands with consideration, he still might feel put upon. After all, he probably had his hands the way he wanted them. He put them there in the first place!

We get no reaction whatsoever. This may mean that Tom is guessing incorrectly into what goes on for Dad; or Tom might be on a right track that will get followed later in the visit. Regardless, Dad is somewhere else at the moment.

Stasche: I search around for somewhere else to navigate down the stream of Dad's inner experience. Since his eyes are focused on the far corner of the ceiling, I try to facilitate a possible visual experience: *Yeah, if you're seeing something, Dad, go ahead and look at it.*

Tom: *A quick jerk on the left hand* ~ Indicates possible positive feedback to seeing something.

Stasche: *Yes, and there's some small eyelid movements, indicating a possible inner visual process. Visual, try for it . . .*

LOUD BEEPING from the stomach tube feeding machine near the next bed. It BEEPS nine times.

Stanley: A very brief pause in chewing movements. We can't tell for sure what this pause may be a reaction to, our interactions or the beeping machine.

Stasche: *Now the strongest thing that I notice is actually the posture. ~ Let me put my hand just on the back of your head here, Dad. Oh yeah, it's like rigid. Shoulders are up off the pillow, your back is up.*

Tom: *So that's why I said, "I'm feeling rigid." When we took both hands it got more rigid.* Tom picked up Stanley's rigid posture signals unconsciously and felt them as rigidity in his own body.

Stasche: *The left hand grabbed harder. Yours too?*

Tom: *Yeah.* Tom has been holding Dad's right hand.

Twenty-two more BEEPS. Tom and I ignore the noise and continue talking. Dad ignores the beeps too.

Tom: *Good strong grip here. Released immediately. I really like the way you shook my hand, too.*

Stanley: Smiles slightly. We're getting communication going.

Tom: *He did that grip right now, too.*

We are "fishing" for where the momentary energy of interaction lies. Dad does not react to this last series of obnoxious beeps, but he squeezes his hand again as we talk quietly about his hands.

Stasche: *Good deal.*

Stanley: Moves his mouth a bit, then comes out with, *Oh yeah.* Tom and I miss this amidst the beeps. I catch it later on video.

Tom: *I squeeze and he twitches back.*

Stasche: *Oh, I got that one too.* On the other hand. *Right on, yeah!* I cheerlead enthusiastically.

Tom: *Hang on tight!*

Stasche: *I've got to check and see if these hands are getting in there.* ~ In the video frame.

I'm going to let go in a minute. First I'm going to hang on tight, Dad.

This may sound contradictory. The idea is to follow and encourage Dad's holding on process enough so that he feels at least temporarily satisfied with our communication, enough to let go more easily. I try to support both needs, his to hang on and mine to check the video.

As I squeeze moderately hard I yell, *Heeyah! Heeyah! Heeyah! I'm going to slip my hand out.* As I pull away Dad tries to hang onto my hand and turns slightly toward me. Why did I yell? I wanted to convey my appreciation of the power that he held my hand with.

Lots of hand action. Do you want to come with? Yeah, right on!

I disengage and adjust the camera angle.

Six more LOUD BEEPS. Mercifully, the nurses enter and take care of the beeping machine.

As I approach Dad, he opens his hand a bit as if in anticipation of my grasping his hand.

Stasche: *Okay, I'm going to take your left hand again.* I do so and say very gently. *That's beautiful, Dad, beautiful.* Sliding my fingers inside his hand.

Tom speaking under his breath: ~ *His eyes followed you the whole way.* ~

Stasche: *Great! Right on.*

Stanley: Looking into my eyes.

Tom addressing Stanley: *I told you Stasche would have a beard, didn't I? Getting gray too.* ~ In a teasing tone.

Stanley: Nods his head slightly several times. Tom and I break out laughing.

Tom: *Pretty soon he'll be an old goat like you.*

Stanley: Moves his mouth a bit and utters a quiet, *Yah.*

Stasche: *Yeah, there's something more you're going to say there.* We pause for response time. *Yeah, now I want to go work with that chewing. So, I'm going to put my finger on the bottom of your chin; here we go. Right now, yeah.* I place my finger gently on the bottom of Dad's chin and push up slightly. He stops chewing.

Stasche: *Wup, where'd it go?*

Stanley: Resumes chewing with my finger in place. I push up against Dad's jaw because it pushes down. I'm trying to help him become more aware of what's going on in his jaw. Jaws use very powerful muscles with many functions like eating and talking and kissing. I'm still fishing, hoping he'll go on talking, since he made a "yah" sound.

Stasche: *Yeah . . . Mmmhmm.*

Stanley: Halts his chewing and sighs really big.

Stasche: *Oh, big sigh, good.*

Stanley: Resumes chewing.

A death in the family

Stasche: I remember some family news and decide to trust that the thought appeared during this visit because it may be something that I need to say and Dad would want to hear. *You know, Fran's sister died this week. Mmmhmm.*

Stanley: At first chewing and staring off into space. Then his mouth moves slowly and eyelids blink. Dad then turns his head directly toward me, makes eye contact and "speaks:" *Haaaagh.*

Stasche: *Yeah, she did. She did. It's too bad.*

Stanley: Dad's eyes look away and head returns to center.

Stasche: *But she had cancer really bad and it was time.*

Stanley weakly: *Beyaa heyaa.*

Stasche: *I'm sorry, Bernice died.* Perhaps Dad had tried to say "Bernice." *She had cancer of the liver, and it was good that she went. She was having a tough time.* Long pause to give time for feelings and other reactions. Indicated by his eyes looking up, Dad probably is envisioning things to himself. Bernice and her family were good friends with our family, even lived next door for awhile when I was a baby. This is Dad's biggest reaction so far this visit. Now we "converse."

Conversing

Stasche: *Yeah, mmmhmmm, there.* I touch the bottom of his chin again.

Stanley: *Haghhr.* ~ Very softly from the back of his throat.

Stasche: *Mmmhmm.* ~ As I move my head closer to hear.

Stanley: More rapid chewing, *Haarghh m mm.*

Stasche: *Haughhh, Haauughughhehh!* I talk back with my version of Dad's sounds to converse with him in his own "language." He could be saying something quite intelligible sounding from the inside, but might need reflective feedback from outside to refine his sounds. I make my sounds longer and louder to demonstrate my enthusiasm and support his talk.

Stanley: *Yeahh.*

Stasche: *Yeaahhh.*

Stanley: Swallows. A swallow can indicate a thought or emotional reaction coming up and then getting swallowed back down again.

Making a comment immediately after the swallow can help the other person catch and hold onto what came up.

Stasche: *Yeah, there was another one, feeling, or thought. Mmm hmm.*

Stanley: *Hughhhh.* This longer response indicates positive feedback to supporting the feelings and thoughts coming up for Dad.

Whose emotions?

But I crump out and avoid the unfolding communication by fiddling with the microphone lying next to Dad's head on the pillow. I reach the limit of my own ability to go into my feelings about the death of Aunt Bernice. Dad's reactions are triggering my own, or perhaps more accurately, we are triggering each other.

Being emotionally involved and trying to maintain my "observer position" is always very difficult for me when working with Dad. The emotions that Dad would touch and then retreat from are often the same ones that I back away from, especially the outward expression of these emotions. I believe that this is why Dad's seizures, which he experienced periodically after his cerebral hemorrhage in 1975, had an emotional quality. Dad's neurosurgeon gave Dad's brain a clean bill of health at the time. But Dad might have needed an altered state, the altered state of an emotional seizure to express some of the strong love, anger, grief, fear, frustration, and awe inside him.

Tom riding alongside me on this Alzheimer's river is immensely helpful. He stands more outside the family situation and often maintains a sense of perspective and humor when I lose awareness in my own feelings.

Why does Dad seem to turn away from expressing emotions outwardly with me and then retreat into himself? I work on myself to trust he is doing the right thing. I believe he needs time and inner space to think and visualize and feel for himself before he can come out to interact with me. The danger is that he gets stuck in there. This may be a part of many people's Alzheimer's/dementia process: beginning to go inside and then getting stuck there. The need to retreat from outer reality and complete the building of an inner reality seems a universal, though often neglected process. Everyone at times goes inward and digests information at their own speed and in their own style. Do Alzheimer's/dementia patients have a greater need? Do they suffer from having neglected some parts of their inner life for a long time? Do they get jammed between perception and expression, like all of us do at times? (Randy Buckner, PhD www.hhmi.org/news/buckner_bio.html)

As a facilitator I support people in all their states, and appropriately bring in my own reactions as potentially useful to clients and society. This is my duty and burden and joy as someone who sits with people in remote states of consciousness. Anything less than my best attempt at this would relegate Dad and his last seven Alzheimer's years to the backwater of humanity.

Stasche, after finishing with the microphone: *Mmmhmm, yeah!* A little too cheerily for the subject of death and grief.

Stanley: Sighs. *Heyou sinkmeeyah.*

Stasche: *You're taking a look around.* I go away from talking with Dad here and bring up visual awareness, though we are plainly on a roll with "words." Did I let him or myself off the hook, or both of us?

Stasche, speaking in a soothing fashion: *Mmm hmmmm. I'm going to put my hand on your left shoulder. There it comes. Mm hmmm, yeah.* Pause for reactions. Dad looks farther away from me and then moves his eyes around. *There, that, just go away a little bit. Notice what's happening. Yeah, looking around. . . Some people look away when they go around . . . Ah huh. Ahuh. Yeah. Mmhmm.* I use the third party "some people" as opposed to "you" to intrude as little as possible on Dad's inner space. I want to facilitate what transpires for him, not pull him out too quickly by trying to relate personally with him. If we were making eye contact and relating more, "you" would have been more on track. *Mmhmm.*

Stanley: *Mmhmaya ayaha.* Dad brings us back to the verbal.

Stasche: *Yeah, say that one again. Your jaw's going, now your jaw has stopped. Whew. Mmhmmm.*

Tom: *He looked towards you.*

Stasche: *Did he?* ~ With my ear by Dad's mouth, I'm located too close to see his eyes.

Stanley: *Awistenaw.*

Stasche: *Awistenawww. Awistenawwww.*

Stanley: Swallows.

Stasche: *Yeah, that one there. Hang onto that one too. Important things coming up here.* Long pause. *Mmm mmmm. Yeah. ~ I'm starting to lose my awareness here. ~ Why don't you go ahead and do something, Tom.* During the pause I started to trance out and "join" Dad in his semicomatose state. This is fairly easy to do when receiving minimal feedback during an interaction with someone. I hit the limit of my abilities to relate and start to focus more on my inner thoughts and feelings, thereby losing some of my outer awareness.

Tom: *Okay.*

Stasche to Tom: ~ *Let me grab this hand again. Change your fingers.* ~ Tom, sitting on Dad's right, has been holding Dad's left hand. Tom pulls his fingers out as I slip mine into Dad's left hand. To Dad: *You got it. You got it. Hang on, whew, there. Yeah, whew!*

I talk emphatically, trying to bring myself out of the trance I am heading into. I have spent other summer days blissed out with Dad on the veranda of the nursing home, breathing the hot humid air, smelling the corn fields on the breeze, listening to the cicadas and daydreaming of old times and feelings. Today I want to stay alert and relate, to learn to communicate better with Dad, to help our family and my relationship with Dad.

Tom: *Mmhmmm. I like the way you hold my hand. Your hands are nice and warm today.* No reaction. *I like that chewing.* The chewing changes frequency, indicating positive response. *Chewing over some good stuff there. Now I know you're avoiding looking at me. It's okay to look at me if you want to sneak a peek.*

Stanley: Turns his head and looks at Tom and then nods fairly vigorously. Chuckles and laughs. Tom has cut through Dad's withdrawn state with the direct personal approach!

Stanley: *Isfoayou . . . issaya.*

Tom: *It's nice to see you, good swallow.*

A menu from Tom's restaurant

Tom goes on a fishing expedition for some content of what's transpiring in Dad**.** *You chewing on stuff? What you chewing on? Chewing over some thoughts . . . memories? . . . good memories. Are you going way back? You thinking about Fran? How bout your mother? Mmmm, thinking about Staschie.*

Stasche: *Pause there.*

Stanley: Starts chewing again and grips both Tom's and my hands. We grip back.

Tom and Stasche: *Uhuh, yeah. Mmhmm, uhuh.*

Stanley, in a whisper: *Yeah.*

Stasche: *Uhuh.*

Tom: *Should we get closer?*

Stasche: *Yeah!*

Stanley: Moves his head sideways toward Tom, possibly in a delayed bodily response to Tom's question about getting closer. We miss this cue in the moment, however, and see it later on the video.

Tom: *Oh yeah, it's a big picture.* We are going for Dad's inner visual experience, since we missed Dad's relationship cue of moving his head toward Tom. *It might be a little scary to look at.*

Stanley: Clears his throat and blinks.

Tom: *I know you're not scared of anything. You're the toughest guy I know.* ~ In a slightly kidding tone.

Stasche: *Sometimes things are scary, though.* Pause. Tom and I attempt to represent what we hypothesize as two opposing viewpoints in Dad's inner world: being brave and looking at something scary, and feeling scared and not looking. By bringing out both sides we hope to help Dad's awareness, so he can decide where he wants to go next.

Stanley: Turns his head and looks at me, then turns back to center. We can hypothesize that I'm the scary thing to look at.

Stasche: *Yeah, they are. You can look this way too, Dad, sneak a peek over here.*

Tom: *Take a look at Stasche.*

Stanley: Small sigh.

Stasche: *Whew. Yeah. Wooee.*

Tom: *Boy, it's tough chewing.* Dad still chews when not otherwise occupied with his mouth. *Chewing on both sides.*

Separation and bonding

Stasche: *Yeah, doing great. I thought of some, I thought of some stuff.*

Stanley: Looks at me briefly.

Stasche: *I was thinking back, back when I left home at 18, and I was thinking I wish I'd had more contact with you, Dad. I left and we didn't communicate. We did some; we did alright. But ah, it wasn't enough for me.*

Stanley: *Swallows.*

Stasche: *Yeah, whatever that thought was hang onto it. Probably wasn't enough contact for you either. I'm sorry, Dad.*

Tom: *I'm feeling sad.*

Stasche in a lowered voice: *Yeah, I'm sorry we didn't have more. Let's get it now.* Long pause. *Mmmhm.*

Stanley: No significant change in Dad's actions since he looked at me.

Tom: ~ *Movement in the feet.* ~

Stanley: Looks over at Tom and smiles briefly then looks straight ahead again. Swallows.

Stasche: *Yeah, there was a good one. Something came up there.*

Stanley: Coughs and looks around, waking from his stupor! He seems more present to his outer environment. Dad looks at our joined hands as he and I move our fingers against each other. Looking at the tape, I have to ask myself: "How much did Tom's and my reticence contribute to the trance state that Dad, and to some extent all three of us were in?" Difficult to answer, but I felt then and still feel we were in it together. In other words, "Whose Alzheimer's is it anyway?" Dad might have been in it the "most", but Tom and I are in a trance too. And having read to this point, you, the reader, may also be experiencing a tendency to trance out because of multiple approach/avoidance signals.

Touch no touch

Stasche: *Yeah, you can sneak a peek over this way if you want, Dad.*

Stanley: Immediately looks my way.

Stasche: *Yeah, hi.*

Stanley: Looks back down at our interacting hands on his lap.

Stasche: *Got shy.*

Tom: *Mmhmm. I feel shy too.*

Stasche: *You too? I'm just shy.* Dad's and my hands pause and then resume interacting even more strongly.

Stasche: *Oh yeah, yeah.* Even more hand activity. *Hoo, hahaa.*

Tom: *It's in those fingers.* Stanley turns his head slightly towards Tom. *Yeah, feel it in those fingers don't you. Those fingers are smart.*

Stanley: Small smile and mouth movements like a silent "yeah".

Stasche: *Yeah, they are.* More silent handwork, with Dad's fingers probing and then his whole hand pulling on mine. Then Dad looks at our hands with a puzzled expression. *You know there's one right there. Look at it, feel it.* Our hands rise off Dad's lap. *Yeah.*

Stanley: Opens his mouth as if to say something.

Arm wrestling

Tom: *Looks like arm wrestling to me.*

Stasche: *Yeah! Yeah! Something wanted to come out there, yeah! Ruh! Yeah!* We're getting into a genuine arm wrestle. *Sometimes when three guys are hanging onto each other's hands like this, things come up. Yeah, and then we move a little more.* Dad increases his pushing pulling movements with my hand. I match his strength without overpowering him, so Dad can sense his strength more fully. *Yeah! That's what three guys will do, yeah.* Dad reduces his muscular exertion and turns the back of my fingers toward him and inspects them. *And sneak a look over this way.*

Stanley: Glances over at Tom, instead.

Stasche: *Oh, not too much though, shy again.*

Tom: *It's okay.*

Stanley: Looks at our hands again, then stares ahead.

Stasche: *It's okay, yeah!*

Tom: *Doing good work.*

Stanley: A strong look of puzzlement enters Dad's face. He looks at me beseechingly.

Tom and Stasche: *Uhuh, yeah, good work, great! Hi. Good work!*

Stanley: *Hi, amowna.* Looking at me very worriedly.

Stasche: *Hi, yeah, just be right here. You got it.*

Stanley: Looking to center and slightly down, relaxing his expression.

Stasche: *Ahah.*

Tom: *Yeah, a little puzzled.* Reflecting Dad's facial expression back.

Stanley: Looking to me then away quickly.

Tom to Stanley: *You have a question.*

Stanley: Looking down and a bit sad. His hand has largely relaxed. Dad appears to be withdrawing inside as he now looks down.

Stasche: *Yeah, yeh, come back here, you!* ~ Delivered with mock ferocity as I pull on Dad's hand and he pulls back on mine.

Stanley: Looks at me briefly and then changes expression and coughs with a phlegmy sound. Head down again.

Get over here!

Stasche: *That's it, yeah; maybe that's a little too much pressure.* I'm afraid of pulling on Dad too hard and driving him back inside. I decide to risk it. *I don't know*! I say this in an angry tone. Then louder: *Just get over here, get over here, get over here, now.* I pull on Dad's hand harder. He responds by pulling back.

Dad had been relating with a lot of eye contact, facial expressions, and some good sound bites. I'm mad that he looks ready to retreat, and this probably one of my last visits for months.

Stanley: *Eheheheh. Wawaheen.* Dad looks at his and my hands, now held quite high, about the level of his chin. <u>*What happened, what happened, here*?</u> Gazing at me even more intently. This is Dad's first sentence of the day!

Tom: *What happened, what happened?* Repeating Stanley's words to benefit Dad's awareness.

Stasche: *Whew! Just notice it, notice what happened. That's it.* I reorganize our grip, more palm to palm. *Whew, feel those feelings, Dad. Hang on with that left hand, wow.* We arm wrestle for a minute with varying intensity. Dad sometimes looks intent, other times distant or puzzled.

Tom to Stanley: *You have a question.*

Stanley: Looks thoughtful.

Stasche: *Sometimes when feelings and thoughts get too much, we just go away for awhile.* Dad's arm still holds a strong tension against mine as he stares ahead.

Growly bear

Fast forward thirty minutes of similar interaction: Dad faces and looks straight ahead, smiling and chewing. He leans over toward me about ten inches. His right hand is grasping Tom's left wrist and his right hand is working Tom's fingers. Dad breathes into his lower abdomen, indicating he could be in touch with body sensations. My left hand is resting on his upper chest, attempting a sense of feeling contact. I have my nose pressed up against the side of his left cheekbone, for contact and so he can feel the vibration of my voice directly through his bones, besides through his ears.

Stasche: *I love you lots.* Gently in a low growl.

Tom: *Mr. Softie.* This is Tom's pet name for a gentle, feeling, sentimental side of Dad that has come up in previous visits. *Mr. Softie's in there.*

Stasche: *Yeah, what the, what the hell, yeah, what the hell.* Dad may be feeling soft, but I feel otherwise and decide to chance bringing this out. *Yeah, what the hell are we doing here anyway, huh? Yeah, what the hell are we doing. Yeah, yeah. Yeahhhha! Nnngggghh! Yeahh! Nerrha! Rrra! Rrrra! Rrrraa!*

Stanley: **Big smile, bigger still**, then a swallow. Positive feedback.

Tom: *Oh, good reaction.*

Stasche: Lots of growling sounds by me.

Stanley: Still grinning, indicating continuing positive feedback.

Stasche: *Ruff, ruff, ruff.*

Tom: *Grizzly bear.*

Stasche: *Rrrruff, nrruff, nrrrghhh, rrrraa. Whoa.* I lift my head and look at Dad's shining face. ~ *Oh yeah, positive feedback.* ~ Laughter. *Alright. Yeah.* Lots more growling with my nose back down on Dad's cheekbone.

Tom to Stanley: *You really like that. He's telling you.*

Stasche: Growling. *Yeah.*

Tom: *Smiling with the mouth and smiling with our cheeks and the eyes are going. We're working our hands.*

Stasche: *Everything's going.*

Tom: *Everything's going. Oh, big smile. Big breath. Oh, good, a chuckle.*

Stasche: Continues wild sounds throughout. What a great release of tension and a lot of fun. Dear reader, if you have the opportunity to make "wild" sounds with a loved one in an institutional setting, to feel less silly and embarrassed it helps to pretend that you are a kid again, playing and free to be expressive.

Hug at last

Fifteen minutes later: Dad has been working over my left hand with both of his. He feels, pulls, pushes, turns my hand over while examining it visually.

Stasche: *Yeah, look up here a little bit* (at my face) *Hi! Dad, hi. Yeah, hands are okay, but there's something attached to them, too, yeah, me! Yeah, hi. There it is again. Hi.* Another brief glance into my eyes.

Tom: *A good look! That was a good look; bet you could do that again.*

Stasche: *Mmm hm. Hi. Yeah, it's Stasche here.*

Stanley: *Mumble . . .* ***get over that way too.*** More words in a row! All the while Dad is actively working my hand.

Stasche: *Mmhmm.*

Stanley: *Mumble.* Dad looks in my eyes briefly. *Mumble.*

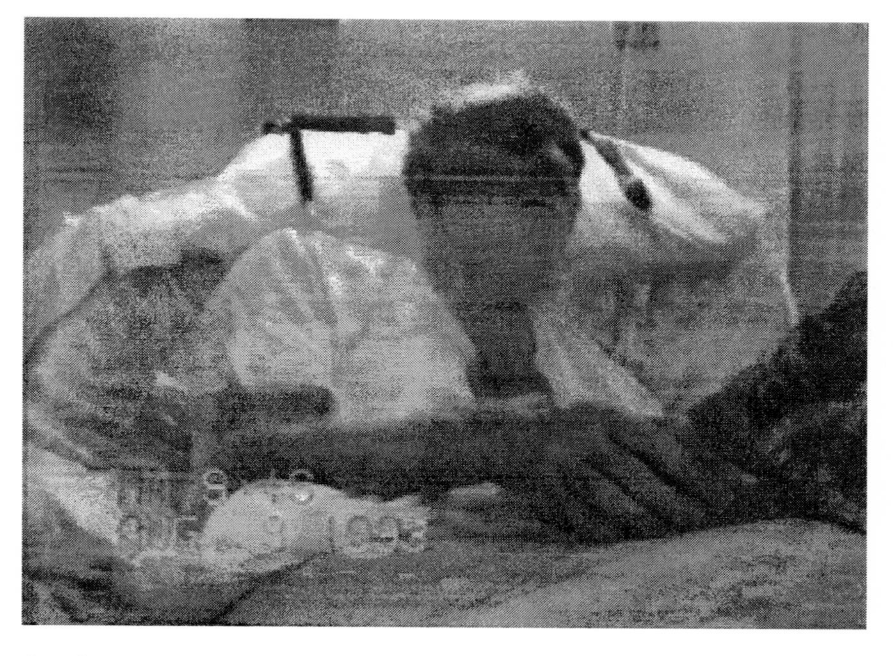

Stanley working over Stasche's left hand while making eye contact.

* * *

Stasche: *Yeah, hi. Yeah, there's someone hitched onto this arm. Yeah, mmhmm.* Dad feels up to my elbow.

Tom: *Work all the way up the shoulder. ~ If you get closer he'll work up your shoulder. ~* Tom encourages us from prior experience.

Stasche: I move closer. *Mmhmm.* Dad works up to my shoulder.

Tom: *Oh, does that feel good.*

Stasche: *Oh yeah, that feels good, that feels really good. Thanks, Dad.* Dad has moved his hands back down to my elbow. I'm wearing a short sleeved shirt. Sometimes he touches my skin and other times he touches the dark blue print. His fingers move rapidly and gently in short movements. He hasn't made eye contact for awhile and seems absorbed in my arm as an object unto itself. *Yeah, do it some more. That's it. Yeah, do it some more. Mmhmm. Yeah.* I place my hand in the center of his upper chest, palm down. He slides my short shirt sleeve up to my shoulder and then lets it slip back down twice.

Tom: *So gentle.*

Stasche: *Mmhmm. Yeahhh.* Dad keeps working. *Mmhmm.* He looks up at my face. *Yeah, hi. Yeah. That's right.*

Stanley: Nods his head emphatically. *Mm oy*. Big smile. Sounds like he said "my boy."

Stasche: *Ha ha ha, oh yeah. Oh boy, that's right.*

Tom: *That's your boy.*

Stasche: *That's true.* Pleased laughter from both Tom and Stasche.

Stanley: Keeps his eyes on my arm he's working with. He now moves his hands a little farther up each time he moves. Tom and I are learning to appreciate Dad's careful pace. Dad glances up at me and then continues his task, reaching up to my shoulder again.

Tom: *He wants you a little closer.*

Stasche: I move a couple inches closer. Looking at the tape, I realize that my hand on Dad's chest makes a certain feeling contact, but also tends to hinder me from moving closer. I think that I'm a little shy to get closer. He seems intent on looking at my arm, so probably our mutual shyness keeps us focused on the arm. *Are you looking at that arm? Go ahead and look at what you're touching. Really notice it.*

Stanley: ***Alright.*** Dad is acknowledging that he has the ability, at least in the moment, to be aware of his communication, to metacommunicate with himself and us! This is an awesome ability for someone in a remote state of altered consciousness.

Stasche: *Yeah, check it out* (the arm and cloth).

Tom: *Maybe we're heading for a hug.* No reaction from Dad to this.

Stasche: *Well, it could be hug time.* I believe a hug to be the right direction, but the timing will unfold of its own accord by supporting Dad and me to relate shyly through touching and looking at my arm.

Get close/back up

Stasche: *Now I just moved, leaned forward and you backed off a little bit.* Dad moved his hand from my shoulder back down to the elbow. I then lean back. *Yeah, now here you come back.* Dad reaches for my shoulder again. This is a get close/back up process. This is all necessary to get to know and feel safe with each other. If Dad knows that I will not pursue him too much when he backs off, then he feels safer to risk getting closer again.

Tom: *You could ask permission for a hug.*

Stasche: *Can I give you a hug, Dad?* No feedback. I'm now sitting on the bed facing Dad. *I'm going to put my hand behind your neck.* I slip my right hand behind Dad's neck. *Yeah, get that in there. Yeah. Oh yeah, that's the one to really work on there.* Dad's right hand works the top of my left shoulder pressing a particular point. *You concentrate on that and I'll sneak a hug in here. Here comes a hug.*

Stasche: *Here comes the hug. Here it comes.* My face approaches his face. Whoa, Dad pushes back slightly on my shoulder. *A little soon. Here it comes.* I move my upper body forward and put my left cheek against Dad's left cheek. His movements remain constant, but a pained puzzled expression comes across his face and remains until I back off.

I'm trying to think how to go farther here. I don't quite know how. Ah haa, yeah. There you go, Dad. Yeah, that's the feeling. I stand up and lean over into a cheek to cheek hug. *Oh boy! Yeah. Mm hmm. Yeah. Yeah. Mm hmm. Yeah. Nice, really nice. Very nice.* Our cheeks remain together. Dad moves his right hand toward my neck and then to my ear and then to the top of my head. His fingers lightly dance over and through my hair. *Yeah. Yeahhh. Mmmm. Mmmm.* These sounds reflect a sense of inner peace, connection, and contentment radiating from deep inside me. *Mm, mmhmm. That feels good.*

Stanley: Dad now strokes the top of my head. The stroking continues to the back of my head.

Tom: *Awwww.*

Stasche: *That's nice.* We continue hugging, with Dad stroking my head. *I love you, Dad. Mm, love you. I love you, love you, love you. Yeah, feel how good it feels. That feels good, really really good. Wow.* In response to my words Dad strokes more firmly and farther down my neck. *That's it. Mmhmm. Yeah, that feels nice.*

Stanley: Slides his hand to my shoulder and pushes me away slightly.

Stasche: I move back and say, *Yeah, hi, oh boy,* as I sit back down on the bed.

Stanley: Searches down my arm with his hand and we settle in a palm to palm grip. My right hand goes to his chest, then to his left shoulder. We look each other in the eyes. Dad starts chewing again, stops and looks at me intently, then smiles.

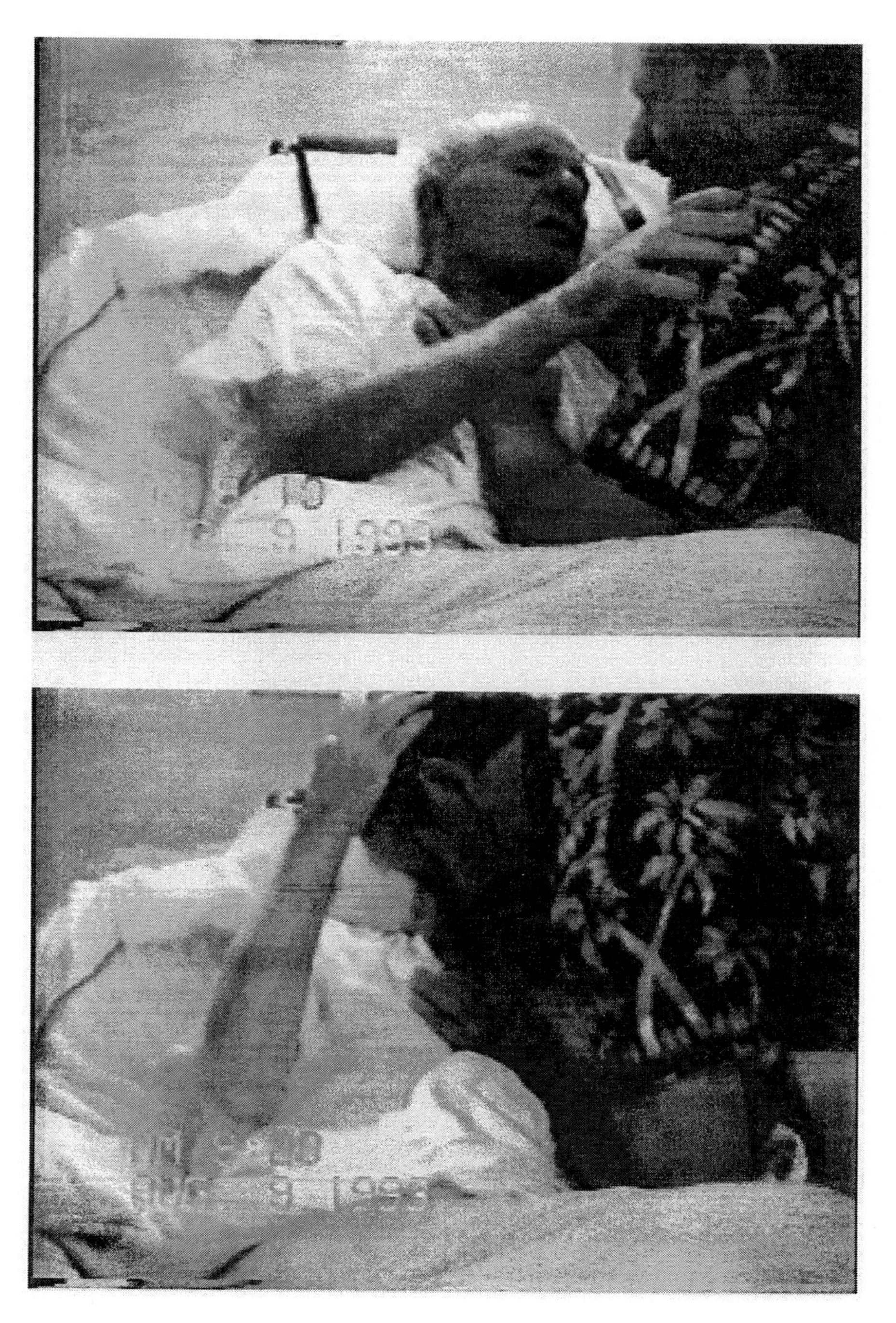

A hug at last. Stanley works his way up Stasche's arm and strokes his hair.

I can hardly write about the emotions I felt then and as I looked at the tape. I feel deep love and devotion, a mystical sense of protection from the difficulties of the universe. He's my Dad; after all, he will always nurture and defend me even after he dies. And now I'm so thankful that we connected in a new and awesome way.

Stasche: *Yeah, hi over here; you can look too. Feelings are good. Touching is good. Looking is good.*

Tom: *Hugging is great!*

Stasche: *Silence is good too, yeah.* Big sigh. We continue to look at each other and hold hands in silence a couple of minutes. We all share in the depth of this experience in quiet communion. You may want to take a moment here to reflect on your own.

Goodbye, goodbye

It's time for us to leave.

Stasche: *Okay, we have to leave soon. Let's start the goodbye . . . I guess.* I'm reluctant to leave Dad. The new emotional connections are a deep feeling treat.

Stanley: *Uhuh.* Big smile. "Goodbye" seems to stimulate a verbal reaction from Dad.

Stasche: *Yeah, okay, so we're going to go soon. We're going to go in six minutes. Yeah, no response, okay. We're going to stay with you a long time, yeah.* Dad grows still and smiles. *And you'll be in our hearts and we'll be in your heart.* Outside a bird starts to sing! *And we're going to come back in a couple of days.* Even though we're not physically staying, I feel the sense of staying with each other in our hearts. *Come back and see you again. You bet.*

Tom to Stanley: *We're gonna' work on that book about you.*

Stasche: *Well, now we're gonna go in four minutes, but we're gonna come back. Absolutely.* We visit a few more minutes and then take our leave.

Stanley: Leans back more relaxed into the bed and pillow. I'm standing; he looks up. I stroke Dad's head and kiss his forehead a couple of times.

Stasche: *Bye bye, see you later alligator.* As I step back, Dad's left hand hangs onto me and he tips his head way back to catch my eyes.

Tom and I drive back to Northbrook, sometime in contented heartfelt silence, and sometime in joyful exuberant celebration. What a great visit!

Consultation and confirmation

Two month's later, at a case supervision seminar with my mentor, Dr. Arnold Mindell, the group reviews a video of this session with Dad. This is the third tape of Dad this group has seen. As in the past, I am vulnerable and sensitive about the condition of my father, my Process Work skills, and my own emotional reactions on the video and in the moment. I also feel proud of the progression of our work and happy for the support of so many good friends. People are getting to know Dad and me better. Colleagues feel touched from watching my work and relationship with Dad.

Arny affirms my ability to maintain enough detachment to perceive the overview of Dad's and my process, and to work effectively with my own father. Arny also confirms that Tom and I are working the right track by supporting Dad's coming closer and going farther away processes.

At the beginning of this chapter I mentioned the patience needed to work with minimal signals and the high degree of sensitivity that can exist behind these signals, how these signals demand new ways of communicating and relating, and how underneath the slow rhythm of Dad's approaches and retreats lies an exciting story of intimacy and connection.

In his pre Alzheimer's existence Dad was a very caring man. He expressed his care as traditional fathers of his generation and culture did: as provider and childhood companion; as protector and warrior (he served in the US Army); as a community leader; and as liaison for our family with the outside world. A traditional view might suggest that Alzheimer's dementia destroyed that caring person. However, Tom's and my experience from working with Dad suggests otherwise. In our opinion, Alzheimer's "transformed" that side of my father's character, my Dad's caring nature, from an outer caring to an inner caring; from a "doing caring" to a "being caring"; from "taking care of people" to "being caring about people, including himself."

There were certainly no handy role models for him to make this kind of transformation, which often caused Dad a lot of pain, recrimination, and guilt. Especially during the early stages of his Alzheimer's he felt incompetent, strange, and weird. By the time of this session he had left these negatives behind. He simply needed support to go farther on his quest for a new style of caring. When Tom and I provided that support, Dad gave us overwhelming positive feedback and love.

During the supervision seminar Arny Mindell said that it is unhealthy for a very caring person to remember everything about what happens to the people they care about. Caring is something that ought to go through a person, not something they should be attached to. Loss of memory forms a way to "let go" of the worries around taking care of people. When we process memory loss, that is, when we go deeper into the new state of consciousness that comes with that loss, we can regain lost parts of ourselves and encounter new aspects of our personalities. At another level from all the suffering, Dad became a more fantastic person. My Dad and I were able to experience a gentle, physical father/son intimacy. The spirit of that intimacy lives with me always.

Chapter Summary

1. **Sensitivity:** Remote, minimally responsive states do not equate with lack of sensitivity. On the contrary, people in these very far out remote states of altered consciousness can be very delicate and sensitive. Communicating with and relating to someone in remote states takes concentrated sensitive awareness of the person in front of us and our own reactions. This is hard work that takes practice; however, starting with a little bit of aware communication can go a long way.

2. **Two or more facilitators:** Having two awareness facilitators is beneficial: a) one facilitator can enter and go deeply into the client's communication process while the other holds the overall awareness, especially if one of the facilitators is a family member who is emotionally involved and more likely to get caught in stuck family patterns and trance out; b) two people can observe more communication signals in more parts of the body; c) they can spell each other; d) two or more facilitators can all cheerlead at once.

3. **Intimacy** with my father is as "scary" for me as it is for Dad and sometimes more so. And I am often less open than Dad, because I am more in my normal identity, and therefore more attached to my ideas about myself than Dad is in his altered Alzheimer's state.

4. **Childhood dream patterns:** Tom focuses on working with Dad's hands because they are important in Dad's earliest remembered childhood dream about the cookie and finger eating bear. These early childhood dream experiences can provide clues to lifelong personality patterns. The growling

sounds that Dad and I make could be related to bear energy patterns. Here's the dream as told by Dad several years earlier: *I am about five years old and somehow have a bag of store bought cookies. Of course we would never have store bought cookies in real life. My mom always baked our cookies on the farm in eastern Nebraska. ~ I take the cookies under the stone bridge over the creek and start eating them. A bear comes along and wants some. I hold out a cookie to it in my right hand. The bear eats the cookie and part of my right thumb and forefinger.*

5. **Patience:** We demonstrate the patience it can take to locate a portal or entry point into Stanley's process. We try everything, including the entire historic menu of things that worked in the past. Every "failure" is another piece of good information as to what does not work in the moment and helps keep us on the right communication path.
6. **Approach and retreat:** The repeating pattern of coming closer and going away, approach and retreat, is typical of all of us, Alzheimer's or not. Two hours may seem like a long time to break through the shyness, but compared to Dad's seventy-eight years of hesitancy around certain kind of intimacy, two hours is short! Dad's hug and stroking of my hair is an incredible moment of intimacy between father and son. This advanced Alzheimer's state has given me a great gift of intimacy with my father, a gift Dad couldn't give and I couldn't have received before Alzheimer's dementia.
7. **The power of withdrawal:** One of the remaining powers that Alzheimer's dementia patients can exercise over their life and circumstances is to come out or go away, retreat further into an inner remote state. If they don't like what is going on,

they will leave. And if they remain angry and withdrawn long enough, they may have a hard time coming back to relate in "normal" ways. Folks in Alzheimer's dementia states are powerful and they get stuck.

8. **Inner life:** Do Alzheimer's dementia folks have a greater need to go inward than they previously did in their earlier, more outgoing lives? Do they suffer from years of neglect of their inner life? Do they get stuck in between perception and expression of thoughts and feelings, like all of us at times? The need to retreat from outer reality and work on an inner reality seems a universal, though often neglected process, for many of us humans.
9. **Who has Alzheimer's dementia here?** This question jumped out at us many times during sessions with Dad and while reviewing videos. If someone with Alzheimer's dementia asks me, "What day is it?" and I answer; and then they ask again a minute later, and I answer; and they ask a third time a minute later; and then the whole cycle repeats a fourth time, well, I have my own case of "dementia" because I seem to have forgotten that I answered four times and the information has not gotten through on a sustained basis! This means I need to: a) find a way to answer the question effectively; b) change the subject to something deeper and more salient; or c) leave the conversation entirely. With Dad we often repeated things that did not garner a response. We then tried something else until we found an interaction that garnered a positive response.
10. **Caregiver incursions:** Caregivers might assume that Alzheimer's dementia patients understand that care procedures must be performed, or that the patients are so remote that the procedures won't bother them. Nothing

could be further from the truth. Please always identify yourself and describe what you are going to do and give your loved one or client a moment to adapt. Notice any response during this moment. If there is no response, proceed with extra care and attention.

11. **Personality transformation:** A traditional view might suggest that Alzheimer's dementia destroyed Dad, that caring person and head of the family. However, our experience from working with Dad suggests otherwise. In our opinion, Alzheimer's transformed that side of Dad's character, his "caring" nature, from an outer caring to an inner caring; from "doing caring" to a "being caring;" from "taking care of people" to "being caring about people, including himself." There were certainly no role models for him to make this kind of transformation, which often caused him a lot of pain, recrimination and guilt. Especially during the early stages of Alzheimer's, he felt incompetent, strange or weird. By the time of this session he had left most of the negatives behind. He still needed support to go farther on his quest for a new style of caring. When we provided that support, Stanley gave us overwhelming positive feedback and love.
12. **Fantastic person:** It is unhealthy for a very caring person to remember everything about what happens to the people they care about. Caring is something that ought to go through you, not something you should be attached to. Loss of memory forms a way to "let go" of the worries around taking care of people. When we process memory loss, that is, when we go deeper into the state of consciousness where we lose our memories, we actually become more fantastic persons in

some ways. As a result of his Alzheimer's, Dad became a more fantastic person. Dad and I were able to experience a gentle, physical intimacy between father and son. The spirit of that intimacy stays with me always.

Chapter 5 Exercise: Being Alzheimer's dementia

Thanks to Ann Jacob for helping create and present this exercise in caregiver seminars.

The point of this exercise is to get in touch with our own forgetfulness to better understand those in Alzheimer's dementia states. Please work to your own comfort level, noting your reactions as you proceed.

1. Imagine a time and place in the future when you have Alzheimer's dementia, perhaps similar to someone you know or have known who is in that condition. Why wait for it to happen to you? Why not find out more about Alzheimer's dementia?
2. Then take a look at yourself in the Alzheimer's dementia state, notice your posture. Listen to yourself, how you would talk or make noises. Feel into how your body would feel. Breathe how you would breathe for a few breaths. Assume the posture you would be in, standing or sitting or lying down. Move how you would move, even just a little bit with one hand.
3. Be there in the Alzheimer's dementia state and notice your attitude about life. Make a note that describes this attitude.
4. Next, still in your Alzheimer's dementia state, forget major things you would like to forget. Just forget them. Let them go. Gone. Poof.
5. Keep a few good thoughts and memories that are important. Write down these good ones.

6. Now, still in you're Alzheimer's dementia state, notice any long forgotten memories or projects, images or feelings that enter your awareness. Note these down along with any new thoughts you have about them.
7. Okay, apply this Alzheimer's dementia state and its accompanying attitude to a career or relationship issue in your daily life. Note how this state would help you sort out what is important in your daily life.
8. How would you view the world differently? How would the world view you differently if you could use your Alzheimer's dementia attitude in the world?
9. Finally, thank your consciousness for at least some forgetfulness.
10. And remember your ability to forget and prioritize is ever ready and waiting for attention!

Exercise notes

Dying has got to be the greatest adventure! Stanley Tomandl

Chapter Six

How Fragile We Are

Fran: *. . . so then after Stanley staples the ugly draperies up on the tavern wall while standing on the dinner table, we leave that party and go to the Klumbs' house. It is 2:00 am or so, and we get them out of bed. We tell them we're hungry. Well we weren't, but they got out some snacks and we played the piano. Then from there we went to Buchers' house. So the four Grohs were with us and our kids, Dan and Stasche. It is now three o'clock in the morning. We demand food. Joyce puts on a baseball mitt that is lying on the table and she says, "I wonder whose dirty old balls were in this mitt?"*

We left their house and picked up some more folks. We drove over to Don's and Campsie's, the tavern owners' house. By that time it was five in the morning. We woke them up. Don came to the door. He had his revolver in hand. He says, "Campsie, will you look in our address book and see if any of these people's names are in there, because they're not my friends." We said we wanted breakfast. Stanley and Emil went down in the basement and got these little steaks out of their freezer. Their four kids woke up. There were about twelve of us by that time. They made breakfast steaks and I don't know what else. And Skip, one of their kids, said, "We never had such a good breakfast!. . ."

* * *

The events that follow in this chapter occur eight months prior to the surprise party.

Stanley enjoying hunting, one of his passionate "meditative" pastimes.

Pneumonia

February 9, 1994: Stanley is fighting a life and death battle with pneumonia and has fallen into a metabolic coma, that is, a coma caused by changes to blood and brain chemistry from his malfunctioning organs. He has been taken to St. Mary's acute care hospital in Kankakee, Illinois.

How fragile Fran sounded on the phone when she called to tell me. Fran: *It threw me . . . it is an emotional thing . . . a rough thing . . .*

February 14: I feel urgent. It has been five days since Fran received the bad news. I offer to take Fran to Kankakee. She jumps at the chance. As we travel down Highway 57 it dawns on me that this will be a Valentine's Day surprise. The question is, what kind of a surprise?

Tom: *So, how's Stanley doing?*

Fran: *When I talked to a nurse this morning she said they weren't going to release him, but they would have Stanley "up" for us.*

Tom: *But as far as his lungs, are they clear yet?* From what I know about pneumonia, I wonder if Stanley has coughed to break up the congestion in his lungs.

Fran: *No, but she said he's doing okay. Susan, our granddaughter who's a nurse, said that once they give him antibiotics, if that's going to work, within 48 hours he'll feel better. And so evidently it's working, but they weren't releasing him today.*

It has been eight days since antibiotics were administered. Fran is relying more on hope than evidence.

Tom: *Oh, when were they thinking of releasing him?*

Fran: *One gal I talked to didn't know.*

Stanley stories

We pass the time on the road telling delightful and deep stories. There are a lot of just plain funny stories about Stanley. But we both know we are not on the way to see the "old Stanley".

Stanley's Alzheimer's dementia

Fran: *Sometimes when I visit, Stanley is kind of sleepy. I don't know if they give him sleeping pills for overnight or not, but by the time I'm ready to leave he's usually okay.* Fran should give herself credit here. Stanley has probably been lifted out of his stupor by Fran relating to him and loving him up.

Tom: *Well, one of the things I learned is that when he was on Three West, before he moved to the New Alzheimer's Unit on Four West, they had him on a drug called Tranxene, which is an antianxiety drug.*

Fran: *That right?*

Tom: . . . *and on Four West, the new Alzheimer's Unit, the first thing they did was take him off it. Apparently by then, there was no reason for him to be on that drug.*

Fran: *Oh, I see. When he was still living at home, I tried to teach Stanley how to shave again and all that personal stuff. That was very difficult. I didn't realize the futility of my efforts soon enough.*

Tom's Grandmothers

Tom: *I don't think you can anticipate how things will go. You gave it a good try. Caretaking someone you love beyond your resources is a perfectly understandable reaction. My grandmother (Gam), who lived to the age of 92, experienced disorientation like Stanley did when he first moved to Manteno, and each time he gets admitted to the acute care hospital. At her personal care facility Gam was moved from one level to the next until she was in total care. With each move she became progressively more disoriented. When she made that last move, she got very disoriented. I think one of the reasons my other Grandmother, Grandma Peg, lived to the ripe old age of 106 is because she lived in the same house she had lived in for the last forty-seven years. She knew where everything was and never became disoriented by disconnection from family and community and the earth she called "Heart's Desire", which she loved to care for and garden in.*

Even when Gam was in full care they would move her from room to room, it seemed like almost every month. When she went to the acute care hospital she'd come back to her personal care facility and be put in a Medicare (government insurance program) *bed. Medicare beds could only be designated to certain rooms. And she'd be in there as long as Medicare was covering it. And then they would have to move her again when the Medicare coverage ran out. Every time she moved she had no idea where she was, you know, completely disoriented. I feel that was part of her dementia problem. When I first saw her at the full care facility, a man walked up to us. He was talking gibberish, so she turned to me after he left and said, "That guy is off in the boondocks, and that's where I'll be soon."*

Fran: *No kidding.*

Tom: *Oh, yeah. So Gam knew that something was coming. My mother brought this up with me. She asked, "What do you do when you know something* (memory loss) *is happening to yourself?" She said, "Gam knew something was happening and I knew something was happening to her." The process makes me sad, but also may contain an element of growth. For something to grow, something else often has to get out of the way. Part of my personal history may have to make room for spiritual and emotional growth. With Gam, one of the things that helped bring her back just after a major stroke was when I said, "Do you remember the boondocks?" She answered, "Yes." I said, "If you want to go to the boondocks, that's okay. But I'll be back at ten o'clock tomorrow if you want to be here, and that's okay too; we can visit." She decided to be there.*

It's important to say to people with dementia, things like, "Oh gosh, I forget things too. I'm sorry you forgot that. Don't feel bad or guilty. That's okay. I forget things too. Why don't we just forget everything right now for a short time and see what's going on."

Fran: *They had an article in the paper the other day about forgetting, and they said people forget where little things are. You write things down, and you still can't find them. Seems like I spent half my life doing that for Stanley. It happens to me, my friends, even young people. When it happens to me, I get frustrated. The article said that's nature's way of not keeping everything up in your head. There are a lot of things you should forget.*

Dying, the most beautiful . . .

Fran: *Stanley used to say, and I don't know how many times he said it, "I think dying will be the most beautiful, glorious thing that ever could happen," and I'd say, "Well, don't rush it."*

Tom: *Yeah! Well, another approach to that would be to say, "Let's try it, but just pretend! How would I die? What would it be like? What would change after I died? What would be different?" What is it that is trying to change that requires something to die? Not necessarily the body. I feel many suicides happen because someone doesn't realize that probably only part of them wants to die. They take death literally and physically rather than as a symbol of personality change that wants to happen.*

Fran: *They want their body to die?*

Tom: *Well, no. They can't differentiate. What's trying to die is usually not the body. They take it out on the body. The body becomes the victim.*

Fran: *Everything is so bad they can't handle it.*

Tom: *Right, and so what's trying to die is outmoded expectations, dreams, beliefs, attachments, ideas, emotions, etc., not necessarily the body.*

Fran: *Sometimes it seems like it's easier to die than to live.*

Tom: *Yes, if you feel like dying, then ask what is supposed to change? You can then pretend to die, to find what is supposed to change. And you may not have to actually die to change. Once, I made a suggestion to the minister of our church who was about to visit someone who had just attempted suicide. Rather than asking what happened, I suggested she ask the person, "If you had been successful, what would have changed?" Our minister reported that it was a very helpful visit: "All this stuff just gushed out!"*

Fran: *So that helped. I can see where your seminars on that stuff would be great! My aunt's husband died twenty years ago. She immediately bought a condo, but never occupied it. I talked to her the other day, and she said, "I'm glad I didn't move. I'm glad I've stayed in my house for the last twenty years." After Stanley moved into the veterans home my sister thought I should move, because I would have to live by myself. I just feel I'm going to live in my house as long as I can. If we were going to move we should have moved when Stanley was still full of life, made a change then, not when he was failing. So I know I'm happier where I'm at. I have all my memories and I have my friends.*

Tom: *Boggles my mind, all this learning.*

Fran: *Stasche is learning, you're learning. I don't know enough about it.*

Tom: *I don't know either. It is like living in two worlds, "ordinary" consensus reality and the world of myth, mystery, imagination and inner adventure . . . having feet in both worlds. Stanley is mostly exploring the second one. Stasche has always been in his own world that way.*

Stanley's medical condition at St. Mary's Hospital

We park at the hospital and make our way inside. Tension mounts as we approach the nurses' station. Fran is vulnerable and hesitant. We are not sure what to expect. We ask where we can find Stanley and for a report:

Tom: *How is he doing? What meds is he getting?*

Nurse: *Antibiotic, for the pneumonia.*

Tom: *Are his lungs clear?*

Nurse: *They are clearing but not clear enough. Since he hasn't been able to do it just by telling him to, we're going to make him cough by suctioning him. It'll break all the mucus up. Okay? Well, it is hard on him. It will force him to cough whether he wants to or not.*

The nurse confirms my worst fear. Stanley isn't coughing. Is this a slow form of "suicide"?

Upon entering Stanley's room, Fran and I receive a big shock. We have never seen him so thin, white as a ghost, weak, lifeless, remote, and fragile.

Wandering around the relationship

Tom: *Hi, Stanislaus!* ~ Stanley's Bohemian birth name.

Stanley: Moaning, gasping, and sighing.

Tom: *Fran's here. Yeah, good swallow. Yeah. She's going for your left hand there.*

Fran: *He's hanging on to my hand here.*

Stanley: *Argh.*

Tom: *You're hanging on. Arghhh!* I reflect Stanley's sounds back.

Fran: *Yeah, he says.*

Tom to Stanley. *Yeah, holding your right hand.*

Fran: *Yeah.*

Stanley: Grumbling.

Fran: ~ *He's hiccupping.*

Stanley: *Ahrrggah.*

Fran: *Are you hiccupping?*

Tom: *Ahrrggahhh, again? What was that?* I am quoting his sounds back to Stanley in hopes of getting a clearer statement. This apparent "nonsense" vocalization requires patience. In a concerted effort to establish communication, I work on the premise that what transpires is meaningful to Stanley and he's attempting to communicate.

Stanley: Hiccup.

Fran/Tom: *He's hiccupping.*

Stanley: Belch.

Tom: Chuckle. *Great!*

Stanley: Hiccup. Burp.

Fran: *Yeah, you don't like to hiccup all the time, do you? No.*

Tom, after a pause: *A little burp there on the end.*

Stanley: Belch.

Tom: *Yeah.*

Stanley: Hiccup. Moaning.

Tom: *I'm going to start working with the movement in his stomach caused by the hiccupping.*

Stanley: Hiccup, belch, hiccup.

Nurse: *Hi!*

Tom: *Hi.*

Nurse: *Mm mmm. If Stanley would just leave it on* (his oxygen mask). *You know what he does? He wiggles his arm and it comes down. Oh, he always gets the hiccups, but he eats good.*

Stanley: *Anchuly.*

Tom: *Anchhuuuly.* I imitate Stanley, but make the sound longer and louder to let him know I like communicating this way and it's okay to do it more. Stanley, however gives negative feedback by remaining silent, dropping this form of communication for now.

Fran: *Wait until I take your hand again.*

Tom: *Hi, buddy. Yeah, take that hand. Yeah, here you go. ~ His stomach is going up and down when he hiccups. ~ Yeah, that's good. I like those hiccups. Yeah, yeah. What do you see today? Mm? Let's look. Look a little closer. Yeah, closing the eyes. Yeah? Yeah, starting to chew. Yeah, look closer. Yeah, mmmm.* Stanley looks up with his eyes under closed lids. This indicates visualizations are probably trying to enter into his awareness.

Valentine's Day

Fran: *I love you, Stanley? I love you. Can you see who you love, who loves you?* Stanley's eyes remain closed, so Fran hints that he should look at her. He may, however, be experiencing her in other ways, like feeling and listening. *I love you, honey. Smack! Smack!* Fran makes kissing noises. *Want to kiss me?*

Stanley: *Ahu.*

Tom: *What?*

Tom, while Fran and Stanley kiss: *He knows how to do that.* Laughter.

Tom: *Make a little love today, uh? Today is Valentine's Day.*

Stanley: *Which is sglables.* Belch . . .

Tom, imitating Stanley: *Yeah, what sglaablesss. Valentine's Day, yeah, your Valentine's here. You can see her. Yeah, you just kissed her.* I use verbal reportage to help Stanley with his mental awareness in addition to his body awareness of kissing. *Yeah, you heard that. Go ahead and listen.* ~ Pause.

Fran: *He's not holding on tight like he usually does. Sometimes you can't get your hand away from him.*

Tom to Stanley: *Yeah. That's good, okay.* ~ Pause. ~ *Hey, every time you hiccup you raise an eyebrow. That's really interesting. I'm going to touch that eyebrow as it goes up here. Yeah, there they went up.*

Fran: *Double header there for you. Both eyebrows go up this time.* This increased eyebrow movement indicates good feedback.

Tom chuckles: *Yes? That's good! Shaking that head. Yeah.*

Fran is here

Stanley mumbles and then speaks very clearly: ***Shake hands. Fran is here and I know that . . . take care*** . . . ~ This is a lot of words for someone in a semicomatose state! Stanley knows that Fran has arthritis in her hands and that they are tender. He may also know that he might be leaving and she'll have to take care.

Tom: *Yeah, shake hands. Yeah, you know Fran's here and you want her to take care.*

Stanley: *Right.* ~ Pause.

Tom: *Yeah? And you said take care.*

Stanley: *Yeah.*

Tom: *Yeah, uh huh.* ~ Pause. ~ *Yeah. Uh huh. Nice smile.* ~ Pause.

Fran: *Smile?*

Tom: *Yeah. Oop, now he's opened his eyes. He's looking.* ~ *Yeah, a little water in your eye, yeah.* This indicates a strong emotional response. I attempt to boost Stanley's awareness, but withhold an interpretation. They could be tears of sadness, joy, frustration, anything. He feels what he needs to feel.

Fran: *He still looks like he has a cold.* Fran kisses Stanley on the forehead. Stanley came out with a personal caring statement for Fran, but she's not ready for it so she changes topics and talks about Stanley rather than to him. In retrospect I could have helped her and Stanley stay with their personal feelings about "taking care."

Tom: *Fran kissed you on the forehead.* Pause. *Yeah, yeah. Okay, I'm going to put my hand on those hiccups. There we go.* I put my hand on Stanley's abdomen where the most movement occurs during hiccups.

Pushing on the hiccups

Tom: Oh, great hiccups! Oh yeah! Yes!

Fran: *That was a big one.*

Tom: I'm pushing against each of the hiccups. I push back gently with the palm of my hand on his abdomen with the same speed and rhythm of his muscle contractions.

Fran: *Oh, that's what made him smile.*

Tom: *Yeah, uh huh. Yeah, feel that, yeah. Ahuh. Yeah, those hiccups are important. Yeah, yeah.*

Hand jive

Fran: . . . *got two hands.*

Tom: *Okay, hang on to Fran.* I include Fran in Stanley's awareness.

Fran: *I don't think I like this one.* Laughs.

Tom: *What? He's holding too tight?*

Fran: *He can hang on that tight.* Laughs. *He can hurt your hand.*

Stanley: *Sure.* Stanley seems to want direct, strong, emotional communication! He squeezes Fran's hand hard enough to hurt it. This must be a difficult situation for her, sitting by her husband's near death bed. It must also be difficult for Stanley. They relate more later in the visit, but they will require almost eight more months to complete their relationship on this earth.

Tom: *Sure? Okay. Yeah, feel those hiccups. Mmhmm.* ~ Pause. ~ *Yeah.* ~ I pause to wait for possible reactions from Stanley. I could have tried to work with Fran's hurt hand and Stanley's reaction to it. But the relationship seemed stalled or beyond my skills as a facilitator. So I go back to Stanley's sensory grounded information in his body.

Stanley: *Okay.*

Tom: *Okay.* ~ Pause. ~ Chuckles. *Mm, yeah! That was good, a double hiccup, yeah!*

Fran: *I can make him laugh.*

Stanley: *Yes, you have.*

Tom: *Yes, it was.* Chuckle.

Fran: Chuckles . . . *Kitchy kitchy . . .*

Tom: *Good. Good, mmmmm, yeah? Yeah, good, double hiccup. Pushing good.* Stanley pushes his abdomen out twice in rapid succession. *Yeah, you feel that.* I answer with movement by pressing on Stanley's abdomen with a similar signal, twice in response in rapid succession.

Okay, I'm going to have to hold you down with two hands here. I'm going to hold you down. Something's pushing here. Yes, stomach. Yeah, pushed again. Your guts. Yeah. Okay. ~ Pause. ~ *What the heck's going on here? Something in the . . . what's going on in those guts? Mm mm.*

Stanley: Hiccup. *Mumble. Oooooh.*

Tom: *Mm mmm.*

Stanley: Belch. *Ummmmm, she's there . . . she's there.*

Tom: *She's there? Yeah? Take a closer look.* I use the word "look" because Stanley's eyes move upward under his eyelids. This upward movement usually indicates a visualization attempting to reach greater awareness.

Stanley: *I guess she's going to be . . .* Hiccup. *Mumble . . . yeah.*

Tom: *I guess she's going to be . . .* I repeat Stanley's words to ensure that he knows what he said and can complete his statement if he wants to. Pause. *Yeah, take a closer look. Mm mmm. Yeah.* Pause.

Mm mmmm, what's coming up? Yup. Uh? Yeah? I'll look closer with you. It's okay to look, mm mmmm. I offer to "look" with Stanley so he feels supported. He may be scared to see something. Pause.

Tom to Fran: *Are you okay?*

Fran: *I'm holding his hand, and I have to watch so he doesn't grab it hard.*

Tom: *Fran, are you okay?* I repeat the question because Fran is looking away, appearing unhappy, discouraged, and distant, beginning to drift into a trance.

Fran: *I'm fine.*

Tom to Stanley: *Okay, yeah.* ~ Pause. ~ *Mm mmm.. Yeah, it's over there. Yeah, I can see it too. It's up there.* I respond to Stanley's eyes roving around the room.

Stanley: *Mumble.*

Tom: *Yeah. Yeah, let's see if we can take a peek, look a little closer. Look at the colors.* Stanley moves his eyes more rapidly around the room, indicating that looking is on the right track.

Uh huh. Yeah! Colors? Texture. Yeah, looks kind of interesting.

Fran: *People on oxygen have to wear these . . . He used to drive an ambulance. He knows about oxygen.* ~ Pause. ~ Naturally Fran wants to connect familiar memories with present events. Especially, Fran wants Stanley to be here now, like the old days when they were first married and he drove ambulance. This seems a natural desire on Valentine's Day. However, she needs to bring her love and emotion more directly into the hospital room at St. Mary's.

Tom: *Mmmm.* ~ Pause. ~ *Yeah?*

Stanley: *Mumble.* I am pressing his abdomen.

Into the beauty

Tom: *Okay, I'm going to press real light on your abdomen. I've been pushing way down during those hiccups. I'm going to go real light on your hiccups. It's okay, you can take a peek at Fran if you want to.*

Fran: *I can't see you with your eyes closed.* Stanley's eyes being open are important to Fran for relating to Stanley.

Tom: *He's taking a peek at you. He's been peeking at you.*

Fran: *There, you opened them. Now you can see me, I know.* Chuckles. *Sure you've got to open your eyes. It's easier to close them, I guess, isn't it?* ~ Pause.

Tom: *Okay, what do you see up there*? Stanley raises his eyebrows. I attempt to bring awareness to his seeing process. *I'm going to touch your eyebrows here. Yeah, what do we see? Look closer. Yeah, there's some good chewing going on.* ~ Pause. ~ *See if I can find the magic spot here; on the jaw muscles.* In the past Stanley often would say something if I pressed there.

Fran: *He chews now.*

Tom: *Yeah. Mm mmmm. Yeah.* ~ Pause. ~ *Doing good, yeah.*

Stanley speaks very clearly and deliberately: **It's b e a u t i f u l**

. . . it's as beautiful as it can be.

Tom: *It's as beautiful as it can be? Is that it? It's beautiful?*

Stanley: *It is!*

Tom: *It's beautiful!*

Stanley: Pause. *Yahhhhhh!* This "beautiful" from Stanley's remote near death state awes Fran and me. Fran leaps back, mouth open.

During the rest of Stanley's life and after his death, Fran and Stasche and I continue to receive comfort from this experience, knowing that he experienced "The Beauty."

Fran: *Are you beautiful?* No feedback from Stanley.

Tom: *Yeah? It's beautiful. It's beautiful. Okay, let's look a little closer. Yeah.* ~ Pause. ~ *Yeah. Mm mmm. Yeah, you're doing great. Ah ha. Do it some more.* Stanley's body twitches. *Yeah, good one.* Stanley's body twitches again. Stanley looks down, indicating he may be feeling sensations in his body. *Okay, feel it as deeply as you can feel it. Good. Good. There you are, believe in your experiences and know that they will show you the way.* I use a blank access statement, a statement nearly without content, to encourage Stanley to trust his experience.

Kitzy kitzy

Fran doesn't relate directly to Stanley's ecstatic state.

Fran: *Kitzy kitzy. We used to do that to each other.* Chuckles. *Kitzy kitzy.* In her own way, she is attempting to connect to his ecstatic state through love play.

Tom to Stanley: *That's good. Mm mmmm.* ~ Pause. ~ *Mm mmm. Yeah, go ahead, you can look at it. Take a peek. Look as closely as you want. Just one little corner.*

Fran: *Smack! Big kiss!*

Tom: *Yeah? That's your Valentine.* Stanley and Fran now have their eyes locked on each other.

Fran: *He's hanging on again.*

Stanley: *Chuckle. Mumble.*

Fran: *If I could get my hand up here, it wouldn't hurt so much. Oh this is unreal, I can't . . .* Chuckles. *You hold my hand tight, don't you?* ~ Pause. ~ *I won't let you do that.*

Tom: *Hey Stanley, want to hold the hand there? Yeah. That's Richards in your left hand, and Fran in your right hand. Richards . . . Fran's over in your other hand.*

Fran: *Is he holding on tight?*

Tom: *No, he just wiggled it.*

Fran: *If you keep it a certain way it doesn't hurt so much when he squeezes.*

Stanley: Belch. Hiccups.

Tom: *A lot of belching going on.* Hiccups. *Yeah, it's coming up.*

Stanley: *Ahhrrgggghhhh.*

Tom: *That's growling.* To Fran: *You want to growl in his ear? He likes growling.* To Stanley: *Want to growl?*

Fran: *Peek a boo!* Giggles.

Coughing

Stanley: Rattling in the throat.

Tom: *Doing good. Some more. You're doing really good.* I am hoping Stanley will start coughing up mucus. He doesn't, but makes some eye movements. *Go ahead and look.*

Fran: *Now I can feel you squeeze my hand.* ~ Pause.

Stanley: Hiccup.

Tom: *. . . go deeper here. Okay, go ahead. Get all those hiccups. Yeah. Go ahead, look down. Look into those guts.* Stanley looks down at his stomach. *Yeah, feel a good big hiccup. The belching, yeah, feel* *that. Feel those guts moving around. I can feel . . . Oh, that's exciting. Let's do some more. See if you can feel it as much as you want. Feel it some more. Feel it some more.*

Fran: *Kitzy kitzy.* Fran scratches Stanley's chin.

Tom to Stanley: *Yeah, a little more.*

Fran: *Kitzy kitzy.*

Stanley: Hiccup.

Tom: *Good one.*

Fran: *Kitzy kitzy. Kitzy kitzy.*

Tom: *Those hiccups keep coming.* They indicate something more is trying to happen. His eyes are closed, head tilted down indicating that he's feeling sensations and perhaps emotions.

Fran: *Do you know what he's feeling?* This is a major transition point for Fran. She has settled down enough in her own feelings to begin concentrating on Stanley's.

Tom: *No, we don't really need to know. Just by holding his hand, you may help Stanley go deeper into his experience. When he gets what he needs inside, he will pop back out and look or speak.*

Stanley: Hiccups.

Tom: *Mmmm, yeah. Is that right?* Laughs. *Yes, yes, yes! All right. One, two, three. One, two, three. I push down with my palm on Stanley's abdomen with shallow, quick movements, duplicating Stanley's actions.*

Stanley: **Cough! . . . Cough!** Stan coughs very vigorously without suction!

This is a breakthrough. I feel immense relief knowing that now Stanley will probably heal and go back to Manteno Veterans Home.

Stanley: Speaks rapidly in Bohemian.

Tom: *Yeah, good cough. Yeah, you gave me three quick beeps.* Three outward thrusts with his abdomen. *One, two, three. Yeah, good jerk. One, two, three. Yeah. One, two, three. It means something. Yeah, you can hear it. One, two, three. Yeah! One! Two! Ahhh! Good one! One two! One two! One. One. One. Big one. Yeah. Yeah. Yes.* I cheerlead his stomach thrusts that push upward on his lungs and help clear them.

Out of semicoma

Stanley: ***Where are we?*** Indicates he is coming out of his semicomatose state. He's asking about the outer world.

Tom: *We're in the hospital. We're at St. Mary's Hospital. We drove all the way down here to see you. An hour and a half from Northbrook. The big white house, where the big guy lived. So . . . what's up? What are your plans, uh? You looking at eternity, uh?*

Fran: Walks to the window. She looks weak and shaken. Stanley's return to reality, indicated by his coherent question, has given her pause to absorb and reflect on how close she was to losing him. She can now get in touch with her feelings.

He mentioned . . .

Tom: *What?*

Fran: *When he called out "beautiful" before, I thought, uh oh . . . this is it! He's going to die right now!*

Tom: Laughs. *Yeah, that's great. Here comes the big White Light, they talk about in near death experiences.* Peels of laughter. *Nah . . . I'll tell you what. I got a good feeling when he coughed, and started talking gibberish Bohemian. I realized that this wasn't his time.*

To Stanley: *So you're going back to Manteno to see Thyra.* (Stanley's favorite caretaker at the nursing home.) *I think you're going to survive for a while longer, good buddy!*

Back to everyday

Nurse: *Can I feed him?* The nurse's interruption is a signal to me that our work is done for the time being. Stanley has had a big spiritual experience and passed a crisis point by coughing, so now he can get on with everyday things like eating.

Fran: *We are going to leave anyway.* The visit lasts a little longer, then Fran and I say our goodbyes to Stanley and head back to Northbrook and Glenview. The following conversation takes place over lunch in a restaurant on the way home.

Tom: *We were just at my grandmother's* (Gam's) *funeral last Monday. She died at age ninety-two.*

Fran: *Well, you can't live forever. Death is part of life; that's for sure, and that's what I thought about Stanley. Before he got this pneumonia, I thought, people don't usually die from Alzheimer's. He might die. So, I felt emotionally messed up. I thought, well, he really hasn't been here mentally; it might be just as well if he died. And then I had to know how he was doing, and I called the hospital, and they said he was so-so.*

Tom: *Mmmm, he really perked up when he heard Valentine's Day.*

Fran: *Yeah, we used to always make cupcakes and decorate them. And we had special birthdays. We had one of Jeanne's* (Fran and Stanley's daughter) *friends stay with us, and I see her pretty often. She says, "That was a fun place to live." So I feel good about our home.*

Tom: *She lived with you?*

Fran: *She did for a year, not now, about thirty years ago. When she was around Amy's age.* Amy is my twenty year old daughter who rooms with Fran.

As we pull into Fran's drive she says: *I think I better put my Valentine's decorations up, or it'll be too late!*

Stanley used to say when we'd come home from a trip, "Well, this house is still here."

Thank you, thank you a lot, more than you know. Tears and hugs.

Tom: *You're welcome. Happy Valentine's Day!*

Postscript

A few days later Fran sends me a note:

Dear Tom,

When you called and said you would take me to see Stanley, it was like an answer to a prayer. I'm so glad I could see Stanley in the hospital. "How fragile we are" has crossed my mind often since then.

Chapter summary

1. **During this visit Stanley:** 1) gains awareness during remote altered states of consciousness and has a big ecstatic spiritual experience; 2) expresses the desire for strong direct emotional communication with Fran; 3) coughs vigorously to start breaking up the mucus in his lungs; and 4) decides to come back, at least for the time being, to a more normal state when he asks: *Where are we?*
2. **Fran has an entirely different experience.** Fran is in her own altered state from: 1) anticipating with dread that this bout of pneumonia will be the end; 2) fearing that Stanley will squeeze too tight and hurt her hand; 3) desperately seeking the *kitzy kitzy* relationship style and its normality; and 4) experiencing intense anxiety over Stanley's big ecstatic moment, thinking that this is it, the end of his life on earth.
3. **Drugs:** In retrospect I wish we had paid more attention to the drugs that Stanley was on throughout his intermediate and advanced stages of Alzheimer's, and their side effects and interactions. All drugs have side effects and synergistic reactions with each other that vary from drug to drug and individual to individual. Many commonly used drugs have adverse effects on alertness, responsiveness, and memory function. Perhaps lightening up on drugs would have helped bring his awareness level closer to the surface.
4. **Forgetfulness, consciously:** Fran and I discuss typical forgetfulness. Consciously going further into the state of forgetfulness when it occurs can help relieve it by finding how forgetfulness is useful.
5. **Moving trauma:** Any change of living arrangements is a disorienting experience and affects memory loss, Alzheimer's

or not. Moving to Manteno is one such move for Stanley. Going to the acute care hospital is another such move. Stanley comes out of his remote state enough to know to ask: *Where are we?*

6. **Cause of death:** Alzheimer's patients do not die from Alzheimer's. The most common cause of death is pneumonia. Stanley was in a life and death situation. Suctioning could have been used to force Stanley to cough, which is critical to breaking up the mucus in his lungs, so he can recover. However, just as important is the facilitation of his decision about whether he is going to stay or leave this life. Stanley answers "yes" by coughing.
7. **Valentine:** The incredible love and connection, the Valentine's Day phenomenon, between Stanley and Fran is spiritual transcendence in its own right. To have the spirit of their relationship support Stanley's connection to the divine was an awesome experience for us all.

Die before you die, so that when you die, you will not die! The Koran

Chapter 6 Exercise: "Dying"

Deep appreciation to Drs. Arnold and Amy Mindell for developing and teaching many variations of this exercise.

This exercise is to help ascertain what needs to die and what needs to be born in our lives. Why wait until our deathbeds to find out? Work to your own comfort level, and appreciate yourself for the courage to search for new awareness about dying.

1. Get in a comfortable posture. Take a few deep breaths and relax. Then imagine into your "death." How will you die? Cancer, car accident, heat attack, old age, etc. Go ahead and die. Leave your ordinary everyday life behind and travel into the altered state of dying. Trust your experience and find whatever you find. You may encounter God or angels, deceased relatives and friends, go into the "light" or feel deeply into your body, fight titanic battles, remember horrific or delightful childhood scenes, find eternal love or . . . perhaps blessed nothing. Work to your own comfort level. Become thoroughly dead. Take a minute here.
2. Now that you are "dead," go ahead and look back on your living self, your everyday self.
3. Ask what has to go, so that there's room for something else to be born in your life? What activity or addiction, which relationships or perhaps a career, perhaps attitudes or beliefs have to be laid to rest? Note these down.
4. Then ask what glorious new part of yourself, probably one that has been gestating for awhile, what glorious new part is being born? A dream, a new relationship, career, fresh belief or inspiration, etc.

5. Explore that new part and make a quick sketch of it. Just a few lines, a figure, an abstract image, what spontaneously comes.
6. Now go cloudy eyed by squinting at your drawing; let your mind go cloudy, too. From your cloudy state, notice the part of your drawing that fascinates you the most.
7. Then make a movement with one of your hands that matches that part of your drawing. Slow the movement down and repeat it. Stay cloudy. As you do this movement, catch the first thing that flickers into your mind. If it is irrational, that's fine.
8. Become this thing that flickered into your mind. See it, hear it, feel it, move like it, put your body into its posture. Weld its energy into your body. Good work!
9. How can this thing's energy help you live the new part of you that is being born into your daily life, right now today?
10. How will the "reborn" you view the world differently? How will the world view you differently?
11. Thank your death and your spirit for the gift of new life.
12. Rebirth is only a "death" away.

Exercise notes

Chapter Seven

Crisis, Coma, and Consternation

Fran: . . . so after our steak breakfast at Don and Campsie's house we finally get back home. We just get into bed and the door bell rings. Here are the Klumbs. They're getting even with us for waking them up in the middle of the night. This is seven o'clock in the morning. We have to get up and make them breakfast.

This is at the time Stanley had started the North Shore Band. He plays tuba in the band and they have a concert that same night. It was nice, and they also had an added attraction, a modern dance troop. The dancers wore scarves and tights. After the concert a whole group of us went to our house for a party. So, naturally, Joyce and I went upstairs and put on long underwear. We got some old kerchiefs and scarves. We told everybody that we could do that dance, too. So we gave quite a performance. Talk about nuts. We have pictures of Joyce and me in our underwear and scarves. That was some weekend. Stanley and I always had fun; I tell you we did!

Stasche's message

Eight months later, Thursday, September 22, 1994, 7:00 am: a message from Stasche, left on my answering machine, *Hello, T.R., this is Stasche. I just got a call from Mom. She said that Dad is in the infirmary at the Veterans Home. He has a low grade infection with a 103 degree fever, which is high enough for an adult. She picked up the impression from the nurse and doctor that it is quite serious.*

Last night I dreamed about standing on a bluff overlooking a ghost town in the mountains of central Idaho, an eternal place with people who have already died. I was crying in my dream. So I feel Dad is going to leave. This location in Idaho is Stasche's spiritual birthplace, his wilderness heart home for many summers in his teenage years.

Dad's oldest sister Helen just visited too. She felt compelled to fly in from Nebraska even though she has a hard time walking. So she's done that, and my brother was just there. I don't know; a lot of signs point that way. Death. My impulse is to call you and say - I'm pretty emotional (pauses to cry*) - somehow either through prayer or direct intervention, let's help Dad get out of here how he needs too. It's a good chance for him to leave. What to do or how to do that, I'm not sure. I imagine . . . if I was there I would go out and help him process it. I'd tell him to leave. I'd tell him to stay . . . everything . . . boy! Anyway, just thought I'd drop that on you. Sorry, good buddy. Give me a call* *back if you have any great thoughts or ideas or brainstorms. Mom repeatedly cycles around the idea of how he hasn't been at home for over four years anyway. Then she'll take that back a bit by saying she loves him, but mostly Dad's not there with her. I want to find a way to work with her process of loving and leaving. She's going to phone back tonight. Give me a call when you get a chance.*

Stanley's condition

That evening I call the infirmary to get an update on Stanley's condition before returning Stasche's call. The nurse reports: *Stanley is not responsive. His blood pressure is low. His color is bad. He is very pale. He has a low grade temperature. We have him on an IV, oxygen, and antibiotics. He has pain in his abdomen and has been moaning. His respiration is fast. He could have a bladder infection. He may have vomited and inhaled some of it. We are going to do x-rays.*

I say: *I will be there in the morning.* The nurse is pessimistic: *If he makes it through the night!*

I report this information to Stasche and ask him about his state of mind. Stasche says: *I am processing about three hundred pounds of anger, that we didn't "let" him die when he got pneumonia eight months ago. It's painful to go through this again. But we have made good use of that eight months. And he decided to stick around.*

Coma and near death realities

Then we discuss coma and remote state techniques appropriate for this situation. I am concerned that Stanley just might expire while I am working with him. Even if it is right for Stanley and the family and the institution, it is not a pleasant thought for me, a lay person at the time. This is beyond my level of experience.

Stasche: *They almost always don't go when you are working with them.*

Tom: But it has happened to you?

Stasche: *It happened only once.*

Tom: *But it did happen!*

Stasche: *Everything the body does is an effort to grow and transform. Support both sides, the part that wants to leave, and the part that wants to stay. Encourage him to fight for life, if that is in him.*

Off to Manteno

The following morning I set off for Manteno. My tape begins: *Today is Friday, September 23, 1994. It is 10:30 in the morning. I am now outside Manteno Veterans Home going in to visit Stanley. The most amazing thing is that I didn't even call ahead this morning to check out how he is doing. I just took off. What an amazing belief system. I have faith that he is still alive.*

The nurse and the doctor happen to be at the nurses' station when I enter. The doctor reports: *Stanley has aspiration pneumonia in the right lung. Apparently he had a stroke. He can't swallow. He has no gag reflex. He is very sick. He is in a coma. Fran knows.*

I explain that I am going to give Stasche a report to help him decide whether or not to come. The doctor says: *Stasche shouldn't come all that way* (from Portland, OR), *because Stanley can't talk.*

The nurse takes me down to Stanley's room and yells: ***Stanley!*** and shakes him to rouse him. ***Stanley!*** He gives no response. His eyes remain closed. The nurse proceeds to give me a more detailed description of his condition: *His color is better, but only because he is "fever bright." He has a mucousy substance coming out of his rectum* *which makes it impossible to keep in a Tylenol suppository. Blood in his urine returned. He is obviously on the dehydrated side. His urine is very dark, due in part to dehydration. His respiration is better today at 40 compared to 44 yesterday. Yesterday he was almost gasping for air. Today his knees and hands are better. Yesterday his knees were mottled, blue looking and cold, and his hands were cold. The sequence of events was probably stroke, which affected his swallowing, followed by aspiration* (foreign matter introduced into the lungs) *of saliva, followed by vomiting and then aspiration of the vomit. The saliva and vomit in the lungs are causing the pneumonia and fever.*

I thank her for her help and turn off the infernal TV (which I consider a nemesis for Alzheimer's dementia patients). Then I set to work.

Tom: *Hi, Stanley . . . Stanislaus . . . Richards here,* I state in a strong voice. *I'm going to touch your left wrist. Here I come. There. That's Richards. Yeah.*

Stanley: *Hmmm . . .*

Tom: *I heard your hmmmmm . . . Now I am going to touch your forehead. Here I come. That's me touching your forehead . . . Okay buddy. What are we going to do here? . . . Good gurgling sounds coming from your throat here.* I start to gurgle with Stanley to give him feedback: *Arrrgggh.*

The death questions

Okay, I'm going to ask you some questions, and see if you can give me any kind of response here . . . Oh, good swallow! They told me your gag response was gone. I am flabbergasted. The doctor told me Stanley lost his gag reflex because of the stroke and here it is back already! I am really encouraged. I decide not to waste any time getting to the tough questions Stasche had rehearsed me on.

Tom: *Stanley, if you want to die, to leave this planet, move any part of your body.* After approximately eight seconds Stanley clears his throat by coughing. *Okay, I heard your coughing.*

Stanley, if you want to live, move any part of your body. Stanley begins breathing faster after about three seconds. This is mixed feedback leaning more toward living. The time between statement and response is shorter. The goal here is to help Stanley get "unstuck", help him clarify both parts, the part that wants to leave and the part that wants to stay. Once clearer, he can make decisions better.

Stanley, if you want Stasche to come right away for a visit, move any part of your body.

After approximately nine seconds he squeezes his eyes shut. *Okay, you just blinked your eyes.*

Stanley, if you want Stasche to stay home in Portland, to stay away, move any part of your body.

There is no feedback after fourteen seconds. *You did good work. The message I got is you want Stasche to come and visit.*

If you are in pain, blink your eyes for me. After six seconds he gives a big swallow. *That was a good swallow!*

Well, I'm going to be a bore. I'm going to ask you that question again. If you want to live . . . Stanley interrupts the question with immediate feedback. He gives a clearly vocal grunt.

Sounds like "Stump the fighter" is in there. You're a tough old coot!

I'm going to try a question again. If you want your son Stasche to come and visit move any part of your body . . . You are breathing harder . . .

Stanley: *Hmmm.*

Tom: *I heard that too. It's going to take Stasche a couple days . . . maybe three or four days to get here if he is going to come. I'm going to call him in three or four minutes and tell him what I think you told me.*

Trying to confirm everything a third time, I repeat the "Stasche to visit" question and get positive feedback in the way of shoulder movement to Stasche coming. I repeat the life or death questions and get no feedback either way. I conclude: *Stanley, I wore you out! I'm going to go call Stasche.*

Consulting with Stasche

There is one pay phone in the building and it is being used by a resident. I try the next building and same scenario. So I ask the nurse if I can use a house phone and she graciously hands me the phone at the nursing station. But I am feeling very self conscious because the nurse and the doctor are right there. How am I going to tell Stasche his dad is responding and swallowing in front of the doctor? This is contrary to the information on the chart. I feel embarrassed about contradicting him. I pull the cord out as long as it will reach and sit down on the floor to make the call. I tell Stasche my predicament and he says: *Just tell me.* So I talk away. Stasche wrestles with the question of whether to come or not: *Of course Dad wants me to come for a visit. It's the same old dilemma. He could hang around for a couple more months or longer. I am torn. Thanks so much for being there, Tom.*

It is 12:00 pm when I return to Stanley's room to resume the coma work.

Tom: *Hi Stanley. I'm back. Richards here. I'm going to touch your knee. That's me, I'm touching your knee. I talked to Stasche and let him know how you are doing. He told me to tell you Fran is going to be here tomorrow. I took my hand off your knee. Yeah, we got some good chewing going on here. Uhu . . . That's me looking for the magic chewing spot and pushing up on your chin with my finger.* The chewing goes away. I remove my hands and the signal comes back strongly. *Wow! Big chewing . . . good chewing . . . big chewing. I cheerlead. Yea! . . . Good! . . . Good! . . . I'm going to touch your throat.* I touch Stanley's throat, restrict chin movement slightly to bring awareness to his jaw, and touch his cheek where the jaw moves. I'm hoping this will bring awareness to his mouth area and he will talk.

Stanley: *Grahh . . .*

Tom: *I hear growling . . . There is growling in there . . . Are you angry? . . . Growling like a bear!* I connect with his first childhood dream that has a bear in it. . . . *growling like a bear . . . big bites! . . . big bites of that cookie . . . like a bear taking big bites of that cookie!* I begin growling with him. *Grahhh . . . Grrrrahhhhh . . .* I expand and extend his phrases. I am really getting into it when, right in the middle of my best growl, a nurse walks in. Fortunately she knows me.

I tell her: *The bear is here.* I have no idea what she thinks of that statement. She checks on him and leaves.

The bear is home. You are really chewing . . . and now banging those teeth together. I put my finger on his chin to amplify the signal and bring awareness by restricting it gently.

There is water in your eyes, Stanislaus! Water . . . water in your eyes! It's good to feel that. Uh huh! . . . Good. You are doing great! Go ahead. Go ahead with what you need to do! The chewing has stopped momentarily. Stanley seems to be feeling more than moving his jaw, as evidenced by water in the eyes.

I return to the question of living or dying and now support the dying side. *Now let's pretend that you are already dead! . . . What would that be like?* Stanley offers no feedback.

Now let's pretend you are going to go on living! Immediately chewing motion returns. *Yea, you gotta do a lot of eating.*

I repeat the question, *What would it be like to go on living?*

Stanley: Within six seconds his entire body jerks in one large spasm. Then he takes a huge swallow and starts vocalizing: *Huh . . .* The gurgling breathing has changed to guttural sounds.

Tom: *I like the way you chew, just like a hungry bear . . . waiting for those cookies.*

Ice cream and life

At this point I spot a Dixie cup of orange sherbet on his night stand. I open it and start feeding it to him. He eats and swallows. After only a few spoonfuls the nurses spot me feeding him. They pop in to attend to him and sit him up. He opens his eyes. One nurse takes over the feeding. He eats half of the orange sherbet. She is surprised and says she will report to the doctor right away. She won't let him finish it, though, because she is afraid he will throw up and aspirate again.

Buddy, I see you still have your good friends (care aides and nurses) *taking care of you. I'm going to check this shoulder here. I'm going to put my hand on your shoulder.* His right shoulder is off the bed. *I'm going to push forward and see if it will go forward.* No go. *Now I'm going to push back and see if it will go back. No, huh. It is right where it wants to be. That's good.*

I touch his forehead, telling him what I am doing. He coughs a really deep lung clearing cough. *Good cough!* I realize that this is the most important part of the work from a physiological sense, because coughing clears the lungs of congestion. But the cough is so deep I can't tell if he is coughing or getting ready to throw up. I panic and call for the nurse. False alarm. He is fine. So the nurse feeds him some more sherbet. Then Stanley's IV runs out. She says: *Stanley, looks like we have to interrupt your ice cream party again.*

The nurse needs to move his right arm, the IV arm. He resists. She says: *I had forgotten how stubborn Stanley can be. You know Stanley. When he doesn't want to do something he goes stiff. If he doesn't want to move, he's not going to move!* This confirms in my mind that not only has he decided to go on living as indicated by his eating, but his personality has returned in full measure. This means that the stroke induced coma crisis has passed, but it also means that his stubborn personality is back. In my experience, I sometimes find it easier to work with people in extremely withdrawn states than in their more normal states, because normal defenses are down in extremely withdrawn states.

The nurse feeds him the rest of the sherbet. She says: *We've had a lot of sick people all of a sudden.* Before she leaves she tells him he will get pudding for dessert later. I kid her about using it as leverage.

Nurse: *Stanley, you're going to have to move that arm if you want any pudding! There you go! Okay Stanley, I will tell the doctor you did very well!* She leaves.

I touch his left hand, after alerting him: *You did great work, Mr. Bear! You are going to get more ice cream and pudding later! You had a good cough. Maybe you want to rest now. If you want me to leave you alone, move any part of your body.* No feedback, indicating negative feedback. *If you want me to stay a little while longer, move any part of your body.* Negative feedback again. *Okay, I'll stick around awhile anyway.* The thought came to me that this might be my last visit. *Before you go to sleep I want you to know that Fran is coming to see you tomorrow.*

You are a lucky guy with all these good friends (nurses and care aides). *I should be so lucky . . . ice cream . . . pudding . . . friends . . . I should be so lucky!*

Stanley bursts out: ***Haw! Haw!***

Tom: *I heard that. You're talking to me. I'm going to sit down in a comfortable chair.*

Stanley starts talking in sounds: *Awww . . . Uhhh . . .*

I mirror and extend his sounds. He coughs. He is straining his lower stomach with these sounds. *Go ahead and feel what you're feeling.* We grunt together. Then he coughs again.

Stanley: Continues, *Mmmmm . . . Mmmmm . . . Mmmmm . . .* Then puffs out his cheeks by blowing into them, followed by three more coughs.

I experiment with different sounds . . . volume, length, phrasing. His phrases get longer: *Uhhhhh . . . Ohhhohohoh . . .* We continue our conversation in Stanley's sound language.

Another nurse comes in and together, using all of our wits, we coach a dose of Tylenol down his throat. Once he knew it wasn't ice cream, stubborn Stanley would not open his mouth!

She appraises Stanley's condition: *His urine is very concentrated. It is dark orange. That is due in part to him being dehydrated. It will take a couple of days to see if his kidneys will kick back in. That and to ascertain if the antibiotics take effect. Dehydration is not an uncommon thing for those who can't walk and get a drink of water. They don't know to tell you, "I'm thirsty." Or they are not even aware enough to know they are thirsty. Or they are yelling, and you don't know what they want.*

Tom: *Okay buddy, well, I think I'm going to take off here. If there is anything you want to tell me or show me, do it now because I am going to leave . . . You're doing real good, Stump. Stump, the fighter . . . the bear . . . the ice cream cookie man . . .*

Stanley starts making sounds: *Ahhhh . . . Ahhhhh . . .*

Tom: I echo and extend his phrases. *Okay buddy. That's me touching your arm. You are doing great. You are nice and cool. Do more of whatever you need to do. Good job. See you later alligator.*

Tom's anger at Stanley

Stanley keeps talking as I leave. It is 1:00pm. I have been working with him for two and a half hours. I experience many emotions as I leave. I am pooped, I am frustrated, I am jealous, and I am angry!

The most ironic thing after all that work is having nothing to say to Stanley! I stood there saying goodbye while he was still talking. And I couldn't think of another thing to say to him. Or more precisely, I felt things I couldn't bring out in the moment. So there he is babbling away, living the life of Reilly. Being waited on "hand and foot" by his caregivers, savoring his favorite foods with no traffic to fight, no responsibilities, no worries.

And here I am in a state of consternation. Instead of celebrating his recovery from a coma, I am exhausted. I feel frustrated at not discovering a single clue about what his unfinished business here on earth might be. He must be hanging around for some reason. I am "jealous" of his lifestyle and his obvious contentment. I am angry at him for his selfishness and stubbornness. "You stubborn old goat, still doing it your way!" All in all, a difficult and wonderful learning experience for me! I stop for a huge jumbo coke. I feel dehydrated!

Consulting with Max

From Manteno I drive to Milwaukee, Wisconsin where I have the privilege of attending a Process Work seminar led by Max Schupbach and Jytte Vikkelsoe. The seminar includes case consultation. I have never presented a case before, especially among professional psychotherapists and social workers. I am in fear and trepidation. Stasche had encouraged me to present regardless. It has to be obvious to Max that I am scared. Even though I am already in the building waiting, for the session to start, I walk in late, sit outside the circle, and present in the last time slot.

Jytte tells Max they both must leave to catch their plane, but Max nevertheless gives me his full attention. I stumble through a description of Friday's coma work session and Max says: *Good for you for working with him!*

Dehydration and ecstasy

Then I state the question that has been forming in my mind: *It seems to me that underlying all Stanley's recent symptoms is the process of dehydration. It intrigues me especially because it involves an element of choice. He drinks sometimes and not others. Dehydration wasn't identified in any of the medical reports until I got in the room. And then, even though it might be the root cause of everything, including the fever and the stroke, it was never addressed as a possible cause. The nurse mentioned that it happens all the time. Can you tell me something about the process of dehydration?*

Max: *First of all, how did you ask him if he wanted to live or not?*

Tom: I tell Max the questions I used with Stanley.

Max: *With dehydration, the best way to ask the question is to put a little water on his lips. If he wants to live, he will take the water or lick his lips. If he doesn't want to live he will reject the water. Try this several times over several days. It is the only valid test under the circumstances. That is because, if you want to die, dehydration is the best way to do it. It is like a roller coaster ride, a real high. You have visions. Have you ever been really really thirsty? Well, you have hallucinations when you are very dehydrated. In the desert thirsty people see mirages. It is like a rock and roll life!*

Next time before starting the IV, test him by putting water on his lips. He will drink or refuse. The question, "Do you want to live or die?" is a consensus reality question which makes no sense to someone who is on a high. You can test this for yourself. Have someone ask you that question when you are high. It will make no sense to you.

In working with Stanley last Friday before consulting with Max, I eventually did use ice cream the way Max uses water. When Stanley took the ice cream it confirmed Max's teaching. That was, in fact, the moment I knew he was coming back. I had the same experience with my grandmother when she lay comatose after a stroke. The magic of ice cream!

Telephone to heaven

Then Max shared the personal story of his mother who died recently.

Max: *She kept telling the family, I want to call my son; it's urgent. The family would say, "Oh, you want to call George?" She would insist, "No! I want to call Hans!" Hans had died several years earlier. They would not give her the telephone, and kept trying to convince her that she couldn't call Hans because he was dead. But, of course, she could call Hans! She knew she could call Hans! She was trying to make a connection on the other side. Once she had called Hans and made her connection, it was safe for her to travel.*

So enjoy Stanley. He sounds like a great guy! He is on an edge to go. Tell him to feel free to go on his trip. He needs to make a connection with the other side. Then he can travel safely.

This helped me understand that making connections could be part of Stanley's unfinished business and I subsequently include it in his surprise party.

Max adds, tongue in cheek: *Also, you should present yourself to the institution as an Alzheimer's "expert" and insist they prescribe Haagen Dazs ice cream three times a day!*

I am most grateful for Max's insight and Jytte's willingness to miss their plane, committed individuals! Max is one of the original Process Work pioneers. Stanley, Stasche, and I thank you, Max and Jytte.

Tomandl family blues

Sunday evening, September 25. I return from the weekend seminar and call Fran. She has been to Manteno twice over the weekend to see Stanley. Both visits were dismal. Stanley was unresponsive. I tell her about the coma work and that the bear (Stanley) wants cookies, ice cream and other goodies. He wants sweetness and pleasure. She says she might bake him some cookies.

Fran: *He must have some control over it. I wonder why he is doing what he is doing?* I answer: *He is enjoying the trip.*

Monday passes. The death vigil continues.

On Tuesday, September 27, I call Fran. She is grieving and adjusting: *My life remains interesting even without Stanley.* She is debating when to visit. She was mourning the fact that: *You got through to him but I didn't.* I suggest that she not worry about Stanley, but worry about what she needs to do for herself under the circumstances. To get through to Stanley I suggest taking ice cream and cookies. I suggest telling stories about friends Stanley was close to who have already died, to help him make that connection to the other side. And I tell her not to boss him around. She is planning to go to visit him tomorrow, Wednesday the 28th.

"Wake"

I confer with Stasche, who has conferred with his Process Work supervisor Joe Goodbread. Stasche shared stories about all the parties his folks threw. Joe especially focused on the story about Stanley when he was an undertaker on the south side of Chicago. Stanley had to stay overnight at wakes to protect his embalming job on the bodies. Apparently some folks would want to get the corpse up and drink and dance with it. Joe's suggestion: *Throw him a great party before he goes, he won't enjoy it so much after he's gone.* Stasche decides his fondest wish for his father is a wake while still alive, a radical idea and seemingly hopeless wish, considering cultural ideas about end of life, not to mention the current mood of the family.

Wednesday passes and to my surprise and Stasche's surprise we learn Wednesday evening that Fran did not go to Manteno. I begin to feel vague nervous "energy" building around me, which becomes crystal clear when I awake very early, remembering a dream this Thursday morning, September 29.

Tom's dream and urgency

My dream is very specific. *In the dream Stasche calls to tell me his dad died.* From this I believe Stanley will die today. The knowledge that I am working up against that last big deadline gives me a wonderful sense of freedom. Freedom to complete whatever seems to need completing for me and for Stanley and his family.

But I also know that my personal sense of urgency and freedom will be difficult to communicate to the family. After all, the family has been suffering the effects of Stanley's advanced Alzheimer's dementia for the last three and a half years. The family death vigil had begun seven days ago when Stanley went into coma after a stroke. Coma work had helped him decide to come back for awhile, but gave no clue as to how to finish his business.

As the morning progresses, my sense of freedom transforms into cheerfulness and a sense of celebration, which I realize could be offensive to a family in the throes of sadness and loss. I am operating off my own dream and intuition, which could easily be called into question. Even considering my close relationship with the family, what right do I have to push my personal agenda on them?

Stasche and I had discussed this the evening before. From Portland Stasche is doing everything he can to work on family issues, including his own growing anxiety over the use of IV's, antibiotics, and oxygen which are beginning to look like heroic measures, something the family had long ago decided to discourage.

Mission to Manteno

I suggest that I can at least offer to take his mother to visit in the morning. She has not made the one hundred and eighty mile round trip to the Illinois Veterans Home in Manteno in the last three days. Stasche says: *Your mission should you chose to accept it, is to take my reluctant mother to see my scared father. Get some sleep. You are going to need it!*

After my dream early Thursday morning, September 29th, I decide to call Fran at 7:30 am. I fear that an untimely phone call could be unsettling under the circumstances; Fran might think it is Stanley's doctor calling with death news. But I want to call early enough to give her the opportunity to invite anyone else who could get away on short notice. She jumps at the offer: *That would be good.* And adds with dread in her voice: *I know he is waiting for me.* Although she goes through the motions of making invitations, it is apparent that she doesn't want anyone else along. I am getting a clearer picture of the "unfinished business." It is between Fran and Stanley.

On the spur of the moment I pack a sunburst decoration, all the hats I can find, and my guitar. Then I stop at the store at 9:00 am to stock up on wine and champagne. I guess I am in the mood for "An Alzheimer's Surprise Party!"

Chapter Summary

1. **The two state ethic** is prominent in facilitating Stanley's near death process; supporting both his altered state opinions and his normal state opinions about whether to live or die. The facilitator must be grounded in sensory information and centered in their own awareness to be able to support the part of the patient that wants to leave, and also the other part that wants to stay. Surprising to some and not to others will be the idea that we all have these parts in our everyday existence, Alzheimer's or not. When we move from our "normal" state into an altered state our perceptions and responses will often change. So the facilitator has to be able to support all sides in normal states and in altered states, too. Max Schupbach gives an example (pp. 122-123): In a normal state the question, "Do you want to live?" has a straightforward meaning and answer. But if you ask someone who is high or in an ecstatic state, that same question won't make sense. You have to ask the question differently, i.e. using water on the lips to get an answer from the part of the person present in the altered state.
2. **Medical curriculum:** The doctor discourages Stasche from traveling to see his father because his father . . . *won't be able to "talk" to him.* The doctor in this case is missing two important concepts; the existence of multiple sentient communication channels, and the need for closure for both Stasche and Stanley. Medical school curriculum could be greatly enhanced by the addition of more training in communication and relationship skills. A corollary benefit would be a reduction in malpractice suits, and malpractice insurance rates.

3. **Sensory grounded sentient channels of communication** include: a) visual, both outer and inner; b) auditory, both outer and inner; c) body sensation; d) body movement.
4. **Binary communication:** One alternative method of communication which is very important when the patient cannot talk is binary communication. I ask Stanley to answer a yes/no question by moving any part of his body, thus freeing him to use whatever means he has to make his response known. (see Mindell, Amy and Mindell, Arnold, and Tomandl)
5. **Dehydration and ecstasy:** The nurse describes dehydration as common from a physical standpoint because of the patient's incapacity to get a drink, ask for a drink or even recognize that they need a drink. Max Schupbach explains why dehydration is common from a spiritual standpoint. Dehydration is a simple drug free doorway to ecstasy.
6. **Not awkward:** Stanley's left shoulder is raised off the bed and rigid. It looks uncomfortable. I test it. But it is not awkward or uncomfortable for Stanley. It is right where he wants it.
7. **Communication privilege:** Despite Fran's many visits and the nonverbal communication training Stasche and I tried to instill in her, she still regrets that I can get through to Stanley and she can't. She does not realize the distinct advantage I have because I bring no unfinished family business, frozen patterns, or baggage to the bedside. I also have more training in this area than she. I realize my privileges and recognize the responsibilities that go with them: the opportunity to take Fran to Manteno and facilitate a final celebration of Stanley's life in this world.

8. **Attend the death?** Stasche agonizes over whether to jump on a plane to get to his father's deathbed. However, his father's unfinished business is mainly an intimate affair between Fran and Stanley. *The kids always used to interrupt us,* as Fran said later after the surprise party. So Stasche wisely intuits that he should arrive later.

Chapter 7 Exercise: Pacing the breath

Deep regards to Drs. Arnold and Amy Mindell for developing coma work, including this pacing technique.

Friends and relatives often feel helpless when faced with a loved one in remote altered consciousness or coma. Pacing the breath can enhance relating and can be taught in a few minutes. Pacing the breath communicates: *I'm with you; I'm here at your side, at your pace.* Upon trying this technique, most caregivers and family members see or feel a response. Practice this exercise with friends, family, and colleagues, even on yourself.

1. Work in pairs. One person lies or sits down, taking a few breaths and images being in coma. Just relax a little and close your eyes. The other person, the awareness helper, begins interacting with step 2.
2. Knock on the door, even when open, and introduce yourself: *Hello, I'm so and so, and I'm going to sit with you for awhile if that's all right.* Notice feedback, any change, or lack of change from your partner, the one in coma.
3. Sit quietly and observe your partner's cues and your own feeling reactions. Pace your partner's breath first by breathing at the same rate, and then breathing into the same place in your own chest or stomach that they are breathing into. Breathe with them for a few breaths. Continue pacing by speaking only on their out breath.
4. Then say: *In two breaths I'm going to . . . put my hand on your right/left forearm . . . Here comes my hand . . . There.* Notice any reactions: eyelids flickering, sounds, changes in breathing rate or location, twitches, small or large movements, etc. These reactions indicate you are on the right track.

5. Then say: *In two breaths, I am going to gently squeeze your arm at the top of each breath.* Proceed to very gently squeeze at the top of each inhalation and let off as the patient exhales. Notice any reactions, especially upon the very first squeeze. You are now communicating with your partner in the body sensation channel.

6. Again check for cues such as sounds, limb movements, eyelid movements, changes in breathing: rhythm; depth; or placement. Support reactions with blank access techniques. Cheer the person on by exclaiming immediately after their cues: *Mmm! Great! Fantastic!* or similar encouragement. You can also say something more channel specific such as: *See what you are seeing and believe in your experiences. Hear what you are hearing, and know this will show you the way. Feel what you are feeling . . . Move how you are moving . . .* When you get positive feedback, and by that we mean a reaction of any kind, however minimal, a slight change in breathing, a flutter of the eyelids, a skin flush, a groan, a swallow, a muscle twitch, etc. continue encouraging that channel: *Yes, that's it, keep seeing/hearing/feeling/moving, knowing that your experience is for you, showing you the way.*

7. When you feel complete (10 minutes for practice), or tired, or have not gotten any positive feedback for awhile, or need to leave, end your visit with something like: *I've got to leave in a moment, if there is anything else you wish to communicate, please do so now* (pause for reactions). *I'll be back* ________ (state when you will return, or say that you will not return and have enjoyed your time together). *Keep experiencing what you are experiencing. Goodbye.* If the patient's condition

allows and you feel comfortable enough, give them a hug, or a light touch on the arm at this time.

8. Come out of the "coma" and give each other feedback. Remember, you're just trying things out, so all feedback is good feedback for your learning. Share what worked and didn't work.
9. Switch partners and do the exercise again. We find that the "comatose" one learns as much if not more than the awareness helper.

Exercise notes

Editor's note: below we return to the first chapter, which is the end of Stanley's life on earth and the beginning of his new adventure.

Chapter Eight

An Alzheimer's Surprise Party

It begins with the sunburst. As I unfold the colorful party decoration, Fran's eyes light up and the party comes to life. Her seventy-eight year old husband Stanley has been in the infirmary of the Illinois Veterans Home in Manteno, Illinois, for the past seven days under a death vigil. He is dying of complications after a three and a half year bout with advanced Alzheimer's dementia. Stanley is now at the end stage, slipping in and out of semicomatose states. And we are throwing him a surprise party? It had been the furthest thought from Fran's mind. Her prospects for the day had been dismal. She had expected to repeat her visit of three days ago, when she sat watch by her unresponsive husband and helplessly listened to his labored breathing.

Fran: *Yes, a party! Stanley, we are going to have a party!* Stanley startles her by opening his half closed unfocused eyes a little wider and nodding his head slightly. These are minimal signals, very small movements from a physical perspective, but very significant positive feedback signals from a man deep in altered consciousness, close to a coma state. The three of us all agree then. All right! It is party time!

So I put up the sunburst. I have tape with me for the occasion, but decide to hang it on the IV stand. It covers up the medical paraphernalia and makes a silent statement on behalf of the family's wish for no further heroic medical measures.

Fran: *Your sweetheart's here, and I love you. Oh, you're raising your eyebrows! It's too bad we didn't stop to get some wine.*

I joyously surprise them as I reach in my bag: *Not to worry. I have white zinfandel (Fran and Stanley's favorite) and champagne!*

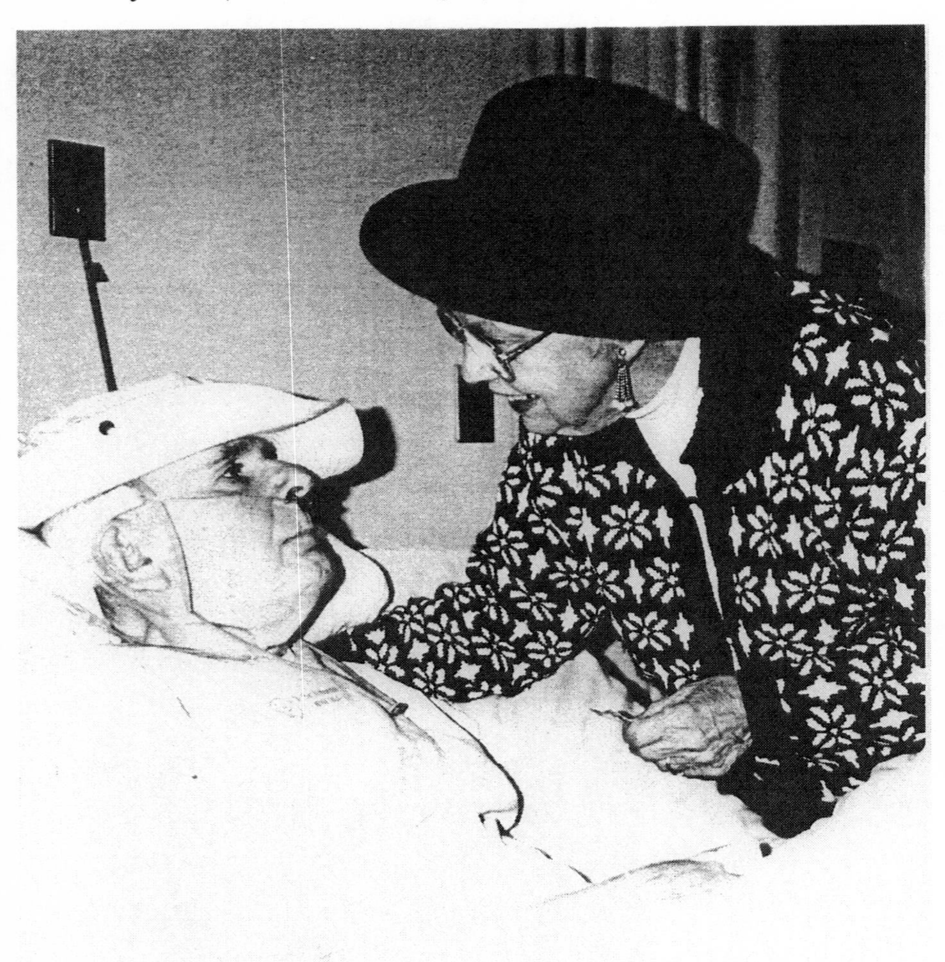

Stanley Tomandl, end stage Alzheimer's patient, completely out of his semicomatose state during his death vigil, relating to wife Francis while enjoying new hats, wine, champagne, ice cream, music, songs, hymns, prayers, intimacy and sacred ceremonial closure at their "Alzheimer's Surprise Party" ~ the day before Stanley dies after three and a half years of advanced Alzheimer's and seven days in a metabolic coma.

Fran: *Stanley, you love parties, don't you.* Stanley immediately coughs, indicating more positive feedback. *Do you want some ice cream?*

Tom: *You've been waiting for your party, haven't you!* Fran and I had stopped at the commissary for double scoops of ice cream. Fran starts feeding Stanley ice cream while I pull out a bag of hats. Stanley has always been a hat man and even wore one in his basement workshop.

Fran: *Oh, hats! He loves hats! What kind of hat do you want today, Stanley?* Fran and I enthusiastically try all the hats on Stanley and each other. We decide on an Australian bush hat for Stanley. Fran says: *It looks just like Stasche's style* (their oldest son, also a "Stanley" or "Stan", nicknamed" Stasche", and co author of this book*).* She dons a stunning black fedora, and I put on a traditional baseball cap.

Fran: *Boy, you look great!* Stanley coughs again. Fran is now in full swing. *Let's get some pictures. What should I do, kiss him?*

Tom: *You can do whatever you want to do. I will take a few pictures.*

Fran gives him a couple of big smackers and exclaims: *I kissed you, Stanley. ~ Look at him look at me! ~ I kissed you honey. Yeah, isn't that sweet. You're a doll!*

Tom: *We're going to write a book about you, Stanley, remember.*

Stanley starts making short sounds: *Uhg . . . Uhg . . .*

I repeat his sounds, making them slightly louder and longer: *Uuhhgg . . . Uuuhhhgggg . . .* This gives Stanley feedback as to what his voice sounds like and helps support him to go farther and modify his vocalizations if he wants to and is able.

Fran: *You're breathing hard. I hope we don't get you too excited.* Fran takes some pictures of Stanley and me and says: *Stasche will like to see you having a party! ~ He sees and everything!*

Tom: ~ *Yes, he looks great. He is completely out of his coma.*

Fran: *We are celebrating lots of stuff.*

Tom: *He is the only guy I know to attend his own going away party.*

Fran: *I think it's never happened before.* Stanley's body is on the last stage of a one-way journey. Fran remembers how Stanley had often told his children that death had to be the "greatest adventure" of all. A celebration seems a fitting send-off for his adventure.

Tom: *You are getting very flushed, Stanley. We are going to celebrate your life.* Fran and I call him by all his names and nicknames; Stanley, Stan, Stanislaus, Stump, and Grandpa.

Now, room decorated, with all of us dressed to party, we break out the champagne. Fran dips her finger in the champagne and touches it to Stanley's lips. He licks his lips, so we all toast his life as he swigs champagne from a paper cup.

Fran: *He is breathing kind of heavy. We excited him too much.* Fran's concern has come up three times now. Stanley is dying, why not let his excitement come up to a certain extent?

Stanley coughs and starts "talking": *Hmmm . . . Hmmmm . . .*

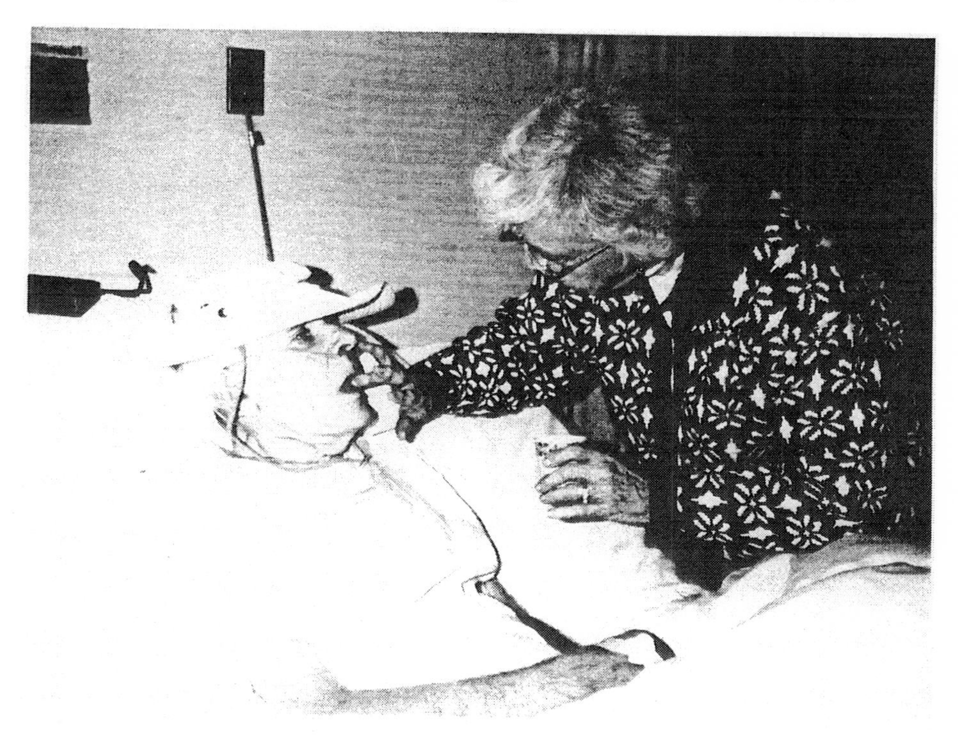

Chaplain ~ Lord's Prayer

At this point I realize it is time for my "chaplain" role and ask Fran her wishes for anything to be included in the celebration. She interprets it as a request being made to the master of ceremonies and defers to Stanley: *He is always the master of ceremonies.*

I clarify my request: *Are there any last rites you want included?*

Fran starts the last rites by reading a newspaper clipping entitled "What Money Can (and Can't) Buy." Between each phrase she hugs and kisses him: ~ *He's looking at me . . . awww. He always said a prayer at lunchtime and we always said the Lord's Prayer together.*

I suggest: *Maybe you'd like to do that now.* We recite an emotional rendition of the Lord's Prayer together and make another champagne toast: *Let's drink to Stanley's life.* Then we take some more pictures.

Fran: *I love you forever. You call me tweedy and I call you tweedy. We're tweedies.*

Stanley coughs; the immediacy and power of his coughing reaction indicates strong positive feedback.

Tom: *He likes that hat.* ~ *This hat has had a life, so I am giving you this hat. This hat is yours to keep.*

Fran: *Look at him raise his eyebrows.* ~ More positive feedback.

Tom: *Well, this is what your son Stasche said he would do if he was here. He'd throw you a party. So we're throwing you a party for Stasche. He's thinking about you. He said he'd throw you a wake. He'd throw you one of those wakes like you attended when you were a mortician.*

Is it okay to take your picture? It is going to be on the cover of your book, you know. ~ *Fran, did you see the feedback? He moved his legs!*

Master of ceremonies

Fran: *He used to call bingo and cheer everybody up at Brandel Care Center. And he used to call square dances. He loved to be the master of ceremonies.*

You know your sister Marie. She was bossy, and she told you what to do just before your were going to do it, and you didn't like that very much.

Tom: *Well, he's broadcasting on all channels. From his breathing signals* (breathing into his chest, diaphragm, and lower stomach all at the same time)*, he is seeing, listening, feeling, and moving. He's on a trip here, a bon voyage trip.* Fran gives him a big gulp of bubbly champagne. I check in with Fran and ask how she is doing.

"In the Mood"

Fran: *I'm okay.*

We're having a party for you, Stanley. Do you know how old you are? You are seventy-eight years old. Kids ask me about how do you know you should get married? I say it is just a natural thing. We just figured we belonged together. We belong together don't we? There is no question about what we did. We could dance and sing to "Elmer's Tune" or "In the Mood." Fran sings the tune.

Stanley coughs and sings: *Oh . . . Ahh . . .*

I cheerlead: *Stanley, you're doing great. You're doing really good. Do whatever you have to do.*

Fran quips: *How about a game of cribbage or gin rummy?* This was a near nightly pre romance ritual in front of their fireplace for years.

Tom: *It looks like Stanley is going inside.* At this point I am calling attention to the change in Stanley's sentient communication signals; eyes closing, stillness in his body, slower deeper breathing, and paler skin tone indicating a deep inner sentient state. Advanced Alzheimer's patients are extremely sensitive and extremely powerful, at least in one way. If they need a break or don't like what is going on, they can stay inside or go farther away.

Fran asks: *Are you leaving us? You're going to go to a different part of the world, huh? I'll see you there sometime. Okay? Jeanne (daughter) and Susan (granddaughter) and Bob and Rita (friends) were all here to see you last weekend.*

Stanley grows very animated: *Ahhhug . . . Ahhhug . . . Ahhhug . . .*

I instruct Fran: *Fran, guess into what he is trying to say.* Fran may be able to intuit what he is trying to say, and then Stanley may be able to confirm her guesses with sounds or movements. Stanley may be clearly saying words in his mind, but he has lost the ability to clearly form the words in his throat and mouth. He may not be able to connect his inside speech to what he is saying out loud.

So I encourage his feedback loop between inner and outer speech. I repeat his phrases and extend them: *Ahhhhhhugu . . . Ahhhhhhuguhhh . . . This is a great trip. A great ride. Pretty exciting. Yea, an exciting ride!*

Fran: *I wish you could tell us about it. I know you said it was beautiful!*

Tom: *We got close to eternity there a couple times.* "Beautiful" and "eternity" refer to words that Stanley uttered at a session nine months previous. He had pneumonia and lay very close to death in a metabolic coma.

Fran: *Yep, a beautiful trip.* We break out the wine and toast his great trip.

Music

Tom: *I learned a new song on my guitar for you, Stanley. My guitar is out in the car and I can get it if you are interested.*

Fran: *Oh yes, Stanley loves music. He loved to play the tuba.* I go out to the car. I leave the tape recorder on. While I'm gone Fran says: *Everybody loves you, Stanley. I love you. Your children love you. Your grandchildren love you. Your great grandchildren love you. Your friends love you. It's nice to be loved by so many people.* From somewhere in the background soft romantic music mysteriously drifts into the room.

When I return, Fran requests polkas, country western, and religious music. We start with a gentle rendition of "Jesus Loves Me." It's just a warm-up. I dedicate the new song I've learned to Stanley and sing "One Moment in Time".

Our party is happening in the middle of a busy institution. A nurse charges in during the song to perform her duties. We keep going and include her in the party. Stanley sings along with me in a strong voice, using sounds instead of words. The chorus about "eternity" is strikingly appropriate to the moment.

The nurse rattles off medical jargon about Stanley's condition, but it feels disconnected, irrelevant at this point. We use her visit to take the opportunity to discuss Thyra and all the other great aides and nurses that have known and loved Stanley at Manteno. This nurse is dressed from head to toe in wild cowgirl regalia. She is real country and western and a member of the Boot Kickers Dance Club. We thank her for her help.

Tom: *This is one of Stasche's favorite country western songs by Ian Tyson, called "Navajo Rug."*

Stanley coughs and sings along.

Fran interrupts the song: ~ *He loves parties. He responded to the song. He is just closing his eyes with the music. Oh, there he opened his eyes as if to say, "Where is it, where is it, where's the music?"*

Another nurse pops in to check on Stanley and encourages us: *What you are doing is better that anything we can do for him now.*

Stanley starts singing his own song and I say: *Okay, I'll sing the last verse.* But Stanley continues to sing his song. I mirror his singing with my vocalizations and extend his phrases. He is using a wide vocal range.

Tom: *I wonder if he is singing in Bohemian?*

Fran: *Helen (Stanley's oldest sister) tried Bohemian last week and he didn't respond at all.*

I quip: *That's because he wasn't singing. Stanley is the one who wants to sing!*

Fran suggests: *How about "Happy Days Are Here Again." Fran sings a rendition:*

Happy days are here again; the skies above are clear again;

we can all afford a Henry Ford; happy days are here again.

Tom: *Oh, I got a good one, "Lonestar."* Fran likes this one and I follow it with "Seven Spanish Angels". Certain phrases from each song catch in my throat.

Stanley starts singing with sounds again.

Fran suggests: *"Hinky dinky parlay voo,"* from "Mademoiselle from Armentières." We give it a try. Then I sing "The Marvelous Toy."

Fran: *Do you know "Hallelujah I'm a Bum?"*

> *Hallelujah I'm a bum, hallelujah bum again, hallelujah give us a handout to revive us again. I don't like work, work don't like me, that is the reason I'm so hungry. Hallelujah I'm a bum, hallelujah bum again, hallelujah give us a handout to revive us again.*

It goes on and on. It's the bum song. It's so familiar because the tune is from a hymn. There was another song, "He Had to Go and Meet her at the Astor." That's as risqué as it got in the 1930's."

Stanley and I sing another duet: "Born at the Right Time."

Fran, going for something lighter and humorous, suggests: *How about 'Dirty Lill Dirty Lill lived on top of garbage hill, never washed and never will?" There is also "Dirty Will." Susan* (granddaughter) *knows the words to that one. The kids used to sing that Tarzan song. At the end of the song it goes: "Jane lost her underwear, me no care, me no care, Tarzan likes her better bare." These were two year old kids singing to Stanley who said, "What are you singing to me?!"*

Fran begins making moves to leave: *Stanley, we had a good visit didn't we? We sure had a good visit. I love you.* But we decide on two more songs: "Kiss the Girl" and the traditional calypso tune "Jamaica Farewell" which ends:

> *My heart is down, my head is turning around,*
>
> *I've got to leave a little girl in Kingston town.*

It's okay

Fran asks me: *Do you know "The Old Rugged Cross?"*

Stanley gives strong vocal and movement feedback.

Tom: *We need some hymns!* Fran and I struggle through "Holy Holy Holy." That is, we struggle with the verses, but we nail the chorus. This hymn is an old friend to individuals brought up in the Western Christian tradition in Stanley's generation:

> *Holy, Holy, Holy, Lord God Almighty*
> *Early in the morning my song shall rise to thee*

Stanley again sings along with even stronger vocalizations, and this time he raises his entire upper body off the bed! This is a man with end-stage Alzheimer's who has not sat up in over a year! There appears to be a deep spiritual connection between his process and the spirituality expressed by the hymn. He also seems to understand the words to the hymn, and may be forming the words in his mind despite his inability to physically express them clearly.

Fran says: *"Rock of Ages" . . . I don't sing it, only when I have to, but I remember it clearly because it was what they sang at my mother's funeral when I was six years old. Most people pick that one. But for him I am going to pick "When We Needed a Neighbor You Were there."*

Tom: *Well, tell him what you are going to do at his funeral.*

Fran: *That's why I like it because it fits our situation. When anybody needed a neighbor you were there. You were always there, Stanley.*

Tom: *Well, we should sing some more hymns. ~ Did Stanley ever mention what he wants at his funeral?*

Fran: *He doesn't sing all those old hymns.*

I softly suggest: *Why don't you climb in bed with him?*

Fran's body jumps at the suggestion, but then freezes: *Stanley, can I climb in bed with you? . . . We can sing the "Doxology."* We stumble through it but pride ourselves on our effort.

Fran kids: *We don't want to sing "Praise the Lord and Pass the Ammunition", do we? How about "Beer Barrel Polka?"* The closest I can come is "Down at the Twist and Shout." *That was a pretty song Tom sang to you. He is a good singer. I bet you thought you were in Heaven already. It's okay if you go on ahead of me."*

Stanley visibly relaxes. The tension flows out of his body and his face softens.

Fran: *He looks so peaceful!*

Stanley as husband, friend, lover, provider, and protector has apparently been waiting for Fran to give him permission to leave, or for her to be ready for him to leave. He needs to know she will be okay. Why is this more than just conjecture? Because of his physical reaction in the moment, and because of Fran's confirmation when I offered to take her on this visit, on the seventh day of the death vigil. She said: *I know he is waiting for me.* In other words, Stanley could have died eight months ago when he was in a comatose state from pneumonia. Or he could have died seven days ago when he was in a second comatose state from stroke and pneumonia. Or he could have died yesterday or the day before. This begs the question, "What is his unfinished business?" This is a salient, deeply touching moment in the surprise party, the profound completion of a significant piece of unfinished business between them, a spiritual healing.

I exclaim: *Already there!? . . . Sorry, we can only walk you up to the Pearly Gates. We gotta check you in with St. Peter. We can probably only wave to Jesus . . . personal savior . . . I wonder who is over there that you are looking forward to seeing?*

This attempt to make a connection to the other side using Stanley's cultural background and spiritual beliefs create a long, thoughtful pause after which we revert back to a sing-along of "This Land Is Your Land."

Bed again

Stanley, my buddy, if you need to go, it is okay with me. If you want to stick around, that's okay, too. But I'll miss you if you take off. I have a premonition that you're leaving today. Last night in my dream,

Stasche told me that Stanley had died. *I think we had a good party. I love you, you stubborn old guy. Your son Stasche loves you and he says goodbye, too. If you have to go, he understands. And if you want to stick around he understands that too . . . Stanley, we gotta go. We gotta go. We will stay around for another ten minutes. If there is anything you need to say, or anything you want us to know, or anything you need to do yourself then do it now, because we have to go in ten minutes. But we'll be here for another ten minutes. Okay?"* I reiterate the ten minute deadline to make sure Stanley has every possible awareness about completing what he needs to complete.

Fran freezes: This is the last ten minutes with her friend, husband and lover on this planet. Then she says: *Tom's pretty good! You were with your grandmother before she died, weren't you, Tom?*

We sing one more hymn, "How Great Thou Art" by Stuart Hine:

O Lord my God! When I in awesome wonder
Consider all the worlds thy hands have made,
I see the stars, I hear the rolling thunder,
Thy power throughout the universe displayed.

Then sings my soul, my Savior God to thee,
How great thou art, how great thou art!
Then sings my soul, my Savior God to thee,
How great thou art, how great thou art!

Stanley, there is water in your eyes. This is evidence that Stanley is experiencing deep strong emotions again, in relation to his spirituality. There is no other physical cause in the moment, such as dust in the air, to account for spontaneous water in both his eyes. I do not try to interpret what these strong emotions might be. This simple observation supports him to feel what he is feeling.

I tell Fran: *I'll put my stuff in the van and give you some private time. Okay? Get into bed if you want to.*

Fran climbs into bed: *Look where I am, Stanley. I'm in your bed!*

I cheerlead: *Now you're talkin'!*

Fran: *I'm in your bed, Stanley. Did you know that? It's your girlfriend. I'm your girlfriend Frances, your sweetie. I'm in your bed and I'm going to hug you.*

Tom: *You can tell everybody you went to bed with him.*

Fran: *It's not as comfortable as I remember beds. They didn't have all this equipment on them. Can I put my head on your pillow? ~ He's got his eyes open a little bit.*

Tom: *Don't worry about the eyes. Do what you need to do . . . Stanley, I've got to go now. I'll see you on the other side. Hope you have a wild ride. Party on, Dude!*

I put my stuff in the van and lay down on the porch bench in the beautiful afternoon sunshine. Fran finds me asleep and says she left Stanley asleep too.

Tom: *That was fun. Stanley is the best audience I ever had!*

Fran: *What about me? I was listening!*

Tom: *Yes, but Stanley was singing along!*

As we climb into the van: *Well, Fran, the last time I saw you, you were in bed with Stanley. How was it?*

Fran laughs: *I didn't get much time because the doctor and the nurse came in. It was just like when the kids were home. Thank you, Tom, for giving me this opportunity to say goodbye to Stanley.*

You know, when I get home and tell this story, nobody is going to believe me!

Twenty-four hours later, Stanley stops breathing in his sleep.

* * *

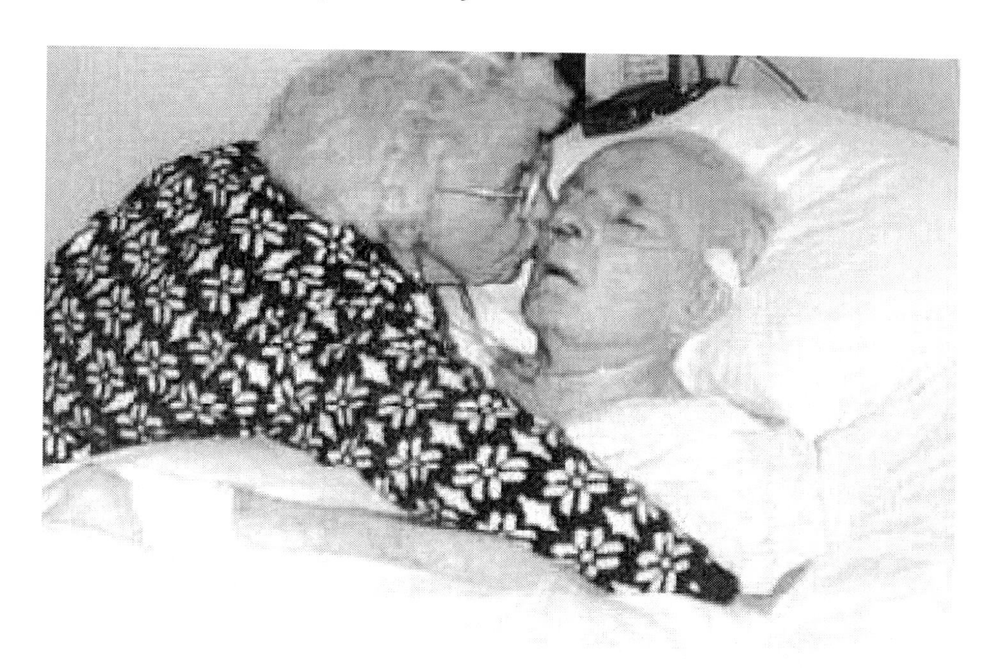

A goodbye kiss as Stanley sleeps at the end of his surprise party.

WHEN I NEEDED A NEIGHBOR

When I needed a neighbor, you were there
I was hungry and thirsty, you were there
I was cold and naked, you were there
I needed shelter, you were there
Wherever I travel, you'll be there
And the creed and the color and the name won't matter
You'll be there

Adapted from Sydney Carter

Chapter summary

1. **A boisterous surprise party** under the circumstances of a death vigil in a Midwestern, USA public institutional setting is a surprising idea in itself. Equally surprising is Stanley's participation. Indeed it is only possible because of Stanley's very responsive positive feedback. Initially he gives small positive feedback signals by opening his eyes a little wider and slightly nodding his head; these amplify into eye contact, vocalizations and relationship; and then into louder melodious vocalizations, significant leg movement, and large upper body movement, none of which have been witnessed for a year. The real "surprise" is the extraordinary extent to which he is present and participating and related throughout the party, despite his end-stage Alzheimer's state.
2. **Observing, commenting on, amplifying, and following** Stanley's communication signals in visual, auditory, body sensation, movement, and relationship channels, I follow his process, providing feedback that he is unable to give to himself. I thus facilitate both inner communication between parts of himself and his outer communication with us. His presence and participation arise from the disciplined application of these feedback techniques, rooted in a positive, supportive, loving attitude.

3. **Words:** Although Stanley cannot formulate words clearly, the words may be clear in his mind, similar to a stroke victim who knows what they want to say but can't get the words out. We can also guess into his vocalizations, and he can confirm our guesses with his positive feedback, like when he wanted more hymns. This is one form of binary (yes/no) communication.

4. **Music** is memorized by multiple parts of the brain and body which makes it a relatively dependable long term memory. For example, piano players can often reaccess forgotten songs by letting their fingers play them. As a musician it is likely that Stanley is actually remembering both melodies and lyrics to many of the songs we sing at his surprise party. For millennia music has been used in the healing arts of many cultures and in the *ars moriendi* (art of dying) practiced in hospices during the Middle Ages in Europe. We feel that Stanley is on his "greatest adventure" and we're still singing with him.

5. **Deep inner journey and relationship:** The surprise party is the last of many occasions when Stanley connects with his deep inner life journey and relates with those around him during his advanced Alzheimer's state. On every occasion during Stanley's advanced Alzheimer's state, if we started our visit by relating to where Stanley was, using sensory grounded sentient communication techniques, as taught by Drs. Amy and Arnold Mindell, and followed his deepest essential processes, he would relate to himself and to us and we would find mystery and intimacy and life and love.

6. **By sentient** we mean being subtly, sensitively, and finely attuned to perceiving minute flickerings of body feelings, movements, sounds, and images that catch your attention. One handy way to access sentient perceptions and sentient communication is to ask yourself, *What am I not noticing right now or almost noticing or just barely noticing?* Sentient perceptions and sentient communications are at the edge of our awareness. It is helpful to welcome them by holding yourself open to thoughts and feelings that are irrational and unexplainable, particularly first impressions.

7. **Sentient awareness** explores the messages, the dreaming, and the meaning contained in sentient perceptions and sentient communications.
8. **Finishing unfinished business** in Stanley's life on this planet, includes assurance from Fran that she will be okay without his physical presence, and her permission for him to stay or leave. This is another "surprise" under conventional ideas about Alzheimer's: the advanced patient is considered incapable of this type of cognitive and emotional processing. People in all sorts of states of altered consciousness, given the awareness and opportunity, often check in with outer reality shortly before they die. The basic philosophy is that people require information from both inner and outer awareness before making major life and death decisions. A piece of Fran's unfinished business is to "let go" of Stanley and assure him that she will be okay if he decides to leave.
9. **Love:** The essential underlying and overarching content of the surprise party, of course, is love between Francis and Stanley, supported by the love of Tom and caregivers at the Veterans Home. The love present in the room is enhanced and communicated by noticing and supporting small nonverbal and verbal signals. This is the love story of one person's leave-taking. Every family will have their own style of expressing love near death.
10. **Creative sacred ceremony:** The surprise party becomes a creative sacred ceremony that offers celebration, blessing, honor, and closure for Fran and Stanley, their family, and friends. This points to the potential for more families suffering from Alzheimer's dementia and other end of life altered states, to experience this kind of deep ceremonial closure and spiritual healing: facilitated, yet spontaneous.

No problem can be solved from the same consciousness that created it.

Albert Einstein

Chapter Nine

Alzheimer's: Beautiful Eternity

Who has Alzheimer's anyway?

Tom: *Life is full of amazing states of consciousness, and one of these states is Alzheimer's dementia. If we try to summarize our research right now; what is it we want to say?*

Stasche: *Alzheimer's, the state of forgetfulness, is one entry point into the mysterious side of life – more accurately of "unlived" life, a different consciousness in a parallel universe.*

Tom: *We ought to say that Alzheimer's folks are really* present *in their own universe, often even more so than they were present in their former normal consensus reality, which most people call everyday life.*

Stasche: *And to the degree we don't succeed in exploring this mystery, this parallel universe, we're going to be more susceptible to Alzheimer's individually and as a society. The disease will force us to explore altered consciousness, if we don't go exploring for ourselves in healthy ways.*

Tom: *When I worked with Stanley, I didn't realize until I actually read the transcripts, how often your dad flipped into normal states, and then I would go into a forgetful Alzheimer's type of consciousness. Who was working with whom?*

Stasche: *I think Alzheimer's was working with both of you. I felt that at times. I had to submit to the spirit of the altered consciousness atmosphere around Dad. Once I did, I relaxed and enjoyed our interactions much more.*

Tom: *Who has Alzheimer's dementia anyway? In my experience, transcripts of normal conversations can read more like conversations between Alzheimer's patients than transcripts of people with Alzheimer's!*

Stasche: *I can't imagine the amount of editing we will have to do on this conversation!*

Tom: *People with Alzheimer's go through stages. The stages are parts of their lives. The fact that the term Alzheimer's didn't even exist a hundred years ago, and the fact that we use it as a catchall makes me wonder. I mean a catchall for anybody who's older and wandering around in an altered state outside of the "normal" universe, or what in Process Work we would call "consensus reality".*

To me the actual breakthrough Alzheimer's afforded your Dad was intimacy of all sorts, emotional and spiritual and physical. He got closer to his own emotions and yours and mine and your mom's, closer to powers greater than himself and closer to you physically.

Stasche: *I would have missed five years of Dad's life without Process Work giving me the courage and tools to open up a new Alzheimer's relationship with him . . . missed five years of beauty and eternity. I mourn his death. He was like some wild lover of the universe, amour de cosmos. He blazed trails through the inner space of his own destiny. We broke through family and culture barriers to find each other with greater son to father and person to person intimacy.*

Tom: *The type of intimacy nearly impossible during Stanley's generation in our Northern Euro-American culture, with its stern, emotionally distant father/son relationship patterns.*

Dying to the consensus reality universe

Tom: *If you have a rigid authoritarian family situation, what is one method of "dying" to escape duties and responsibilities? Answer: forgetfulness. So I see Alzheimer's not only as enforced divorce from reality but also an enforced death from life situations that need to change and open up around human relationship and spiritual relationship; an enforced withdrawal from outmoded ways of fulfilling old roles. That's a heroic thing to do. Now, how consciously do we deal with these withdrawals?*

Stasche: *Well, that's where we come in, in trying to help people, including ourselves, communicate and grow more consciously and more joyfully. I remember the first time Dad didn't remember who I was. That was like dying too. It was a double death; we both got to leave our old identities, we were on our own with each other. It was scary. But, what a gift! A new start.*

What a thorough way to change the context of his life. Dad went totally to another place. During his inner drive for freedom, the context of the Veterans Home changed everything. He fully entered a parallel universe. Occasionally he communicated back to our normal universe, but his interest was over there inside himself.

I remember Mom and Dad's 50th wedding anniversary at the nursing home. How strange and wonderful! Mom was quite a trooper. Talk about her having to stretch and grow!

We bought a large flat cake at the grocery store. The Veterans Home provided ice cream. Residents formed the guest party. Thyra, Dad's favorite care aide, came in on her day off and dressed Dad in a suit. Mom, Jeanne, Ralph, Dan, and I drove out for the celebration. We were all in a context we never thought we'd be in for their 50th. The experience was far from easy for any of us. But now, looking back, I appreciate the richness and courage in the room. We navigated the Alzheimer's universe as best we could, with feelings of sadness and joy, confusion and awe, and much love.

The Alzheimer's journey

Tom: *I tend to bunch everything Stanley did together, saying whatever he did was heroic. Be a hero, go remote. Mythic journey!*

Stasche: *The Alzheimer's journey!*

Tom: *"I'm not going to come out unless you're interested in where I am." It was a kind of heroic attempt to create sacred space for himself and others. And explore the beauty and meaning of life; that's what your Dad enjoyed the most. When I would go down to visit him at Manteno, he would take hold of my hand and then he would go off inside, and twenty minutes later he'd come back. He knew he had permission and safety to go out into extremely remote states. And making spiritual connections, I mean deep meditation, so much of Alzheimer's seems to involve meditation, death, rebirth, and spiritual connection.*

Stasche: *Yes, Dad had a great deep time . . .*

Tom: *. . . he was hungry for more spiritual connection.*

Stasche: *Dad did make a lot of spiritual connections throughout his life, with nature, and within the support of the church. That I know. He was a spiritual Argonaut, I imagine he still is. Certainly he still inspires me to go deeper, quieter, louder, wilder, more outspoken, more soft-spoken, more related, less related; to follow the spirit of the moment with ever more discipline.*

One man's experience can apply very directly for all of us. Everyone can draw their own creative conclusions from Dad's experience and think about their own spiritual life in creative ways.

Normal state

Tom: *We could say the premise of our book is that Alzheimer's is a "normal state". Consider for a moment that Alzheimer's may be a normal state, beyond pathology, and for your father, was a heroic journey that took guts.*

Stasche: *It's a journey, a great journey. Heroic journeys are normal states, and we all journey heroically. We all suffer and sometimes feel downed. At times I'm weak, so was Dad. Other times we rise to meet challenges.*

Tom: *I don't know any hero whom I wouldn't consider a wimp in some respect.*

In every instance, when we went where he was, we would find mystery and intriguing life. He was closer to life than we were. That's why we kept going to see him. We're still working on it. And I'm still angry that he wouldn't teach me more directly about what he experienced. I said, "You're my teacher," and he in effect said, "I'm no teacher; I'm going to disappear if you call me a teacher."

Stasche: *From within the Alzheimer's experience Dad disavowed or was unaware of what he could teach. This matches our consensus reality attitude about Alzheimer's experience. Nothing there. But he did teach us on his own terms, his own stubborn, rebellious, contrary, ornery, loving terms. He taught us about connecting ecstatically with spirit and how to have the freedom to be ourselves in whatever state of consciousness we're experiencing in the moment.*

Tom: *Stanley was 100% present and exploring. In our experience, Alzheimer's patients are all here.*

Stasche: *Alzheimer's is a commonly occurring state. We can all access our versions! This is why we present our experiential exercises throughout the book.*

Plaques and tangles

Tom: *With all due respect to Dr. Alois Alzheimer, I think "plaques and tangles" are overrated as a physical "cause" of Alzheimer's disease. Plaques and tangles were originally discovered during autopsies of brain matter from Alzheimer's patients. More recently, Dr. David Snowdon conducting long term research "found many nuns who remained lucid and vigorous into extreme old age, even though their brains were riddled with diseased nerve cells found in Alzheimer's."* (in Valeo. p. 41) *While I'm not sold on plaques and tangles as a "cause" of Alzheimer's disease, I do think there is a case for plaques and tangles as a "symptom" of Alzheimer's dis-ease.*

Stasche: Also, how about the *symbolism of the word "tangles."*

Tom: *Yes, I have an unusual hypothesis: that the tangles actually exist in other modes, such as relationship and spirituality, before they become Alzheimer's, and may eventually manifest as tangles in the brain. I think the Alzheimer's state is an attempt to more fully enter the tangles of life's experiences. The state is leading to something deeper.*

Stasche: *Continue.*

Tom: *Dr. Bernie Siegel writes in* Love, Medicine, and Miracles, *"We store our childhoods in our bodies. Then one day our body presents us a "bill" in the form of cancer or heart attack or something similar. Well, we could say the same thing, that you tally up a bill, a psychological, emotional, and spiritual bill as well as a medical bill, and in some cases the bill is presented as plaques and tangles, Alzheimer's dementia because we need to go further with our personal, interpersonal, and spiritual development.*

Stasche: *That's great, bills on an inner ledger sheet. No wonder, you were an accountant by profession at one time! Many religious traditions would agree with a spiritual accounting.*

Tom: *That's why Dr. Randy Buckner's research hypothesis: "Dementia may be a consequence of the everyday function of the brain . . . probably a cascade of events that ultimately leads to Alzheimer's"* (www.hhmi.org/news/buckner5.html) *is no surprise to me. Dr. Snowdon also "concludes Alzheimer's must be a disease 'that evolves over decades and interacts with many other factors.'"* (in Valeo, p. 41).

With Alzheimer's you are in a compressed state, lots happening quickly on many levels, accelerated. You're in a speeded up routine, and your speeded up routine could have been cancer, could have been a heart attack or another dis-ease.

Stasche: *Different heroic awareness journeys for different people at different times in their lives.*

Tom: *Right. So we're faced with saying that physical, emotional, mental, and spiritual symptoms are wonderful, normal reactions to, uh, personal and collective problems?*

Stasche: *Normal attempts at growth. The Alzheimer's state is an attempt at growth, at personal growth and also emotional, spiritual, relationship, and cultural growth. Alzheimer's dementia is greater than personal growth and deeper than physical pathology.*

Tom: *We could say it's along the same line as conception, birth, illness, schooling, puberty, love, relationship, menopause, elderhood; all deep and necessary life events and stages. These can be energizing and growth promoting or disastrous, or some combination of the two.*

Stasche: *Exciting and sometimes difficult transitions, that's for damn sure. How about the life school of Alzheimer's?*

Tom: *The Alzheimer's institute of totally useful skills.*

Stasche: *Alzheimer's skills we want to develop more consciously.*

Tom: *Eldership, intimacy, deeper relationship, spirituality, meditation; make your own list. And the extent to which your father grew in his last years was through his Alzheimer's state of consciousness, aided by his own and others' facilitation.*

According to Dr. Randy Buckner's observations "when a person who has clinical Alzheimer's disease is asked to concentrate on a specific task . . . the daydreaming, musing, and thinking to themselves, activity increases in the brain, rather than showing less activity as it would in a young healthy adult." This research corresponds to our experience of Alzheimer's patients going on inner journeys. (www.hhmi.org/news/buckner5.html).

Love story

Tom: *You know this is also your parents' very touching love story. I think there is something to be said about the overall state of love and marriage. I think if you compare Fran and Stanley's relationship to Romeo and Juliet, you could say that humankind has made some progress in the last four hundred years.*

Stasche: *At least we can say there is more communication now, even if some of it has to begin in coma. Altered consciousness can be the quickest way to new information and new ways of communication when an inner or outer facilitator is present.*

Humor and Stanley's generation

Tom: *To what extent was Stanley's humor and joking around edge processing behavior? That is, a way of avoiding* and *approaching difficult scary subjects.*

Stasche: *Well, Dad used humor in all different ways. It was a panacea for him. He used it to gain important information and to work with hard feelings. Other times he penetrated very deeply, very quickly with a good joke or the right remark. Dad could charm people in nasty moods, help folks talk about difficult and embarrassing topics and lighten up heavy situations with his humor. He moved the seemingly unmovable with jokes, stories, and well-timed comments. I remember one time a friend came over for a dinner party at my folks'. Dad could see the man was in a grumpy mood as he entered the front door, so Dad grabbed the guy's hat and threw it out into the street yelling as the guy went to retrieve it, "Come back in when you're cheerful! I want to have a good time tonight!" The friend lightened up and came back in the house.*

Tom: *The other thing I was thinking that should be in the book is that your Dad was a man of his times, protector, provider, family man, community leader . . .*

Stasche: *. . . he represents that generation very well.*

Tom: *That makes me feel sad. He represents the end of an era!*

Stasche: *He was a benevolent disciplinarian as a father, a parental style leader. He had a very independent spirit, yet worked well within organizations. People looked up to him and counted on him to raise and resolve thorny issues. He stood for justice in large and small ways. When I was eleven years old my best friend and I started to gang up on my younger brother. Dad stopped us and explained about the unfairness of ganging up on somebody. He wisely avoided the mistake of directly protecting my brother, which would have made us get revenge later.*

Phases

Tom: *Well, for Stanley's Alzheimer's, I am thinking the first phase involved unfinished business: regret; shame; guilt; unresolved mistakes; anger; injustice; revenge. This led him to meeting and becoming new parts of himself, the inner parts: Mr. Softie; Mr. Intimate; the Reverend; and the Wise Elder. Some parts he projected outside of himself: The Big Guy and Mr. Tough Guy. The middle phase was harvesting a lifetime of experience. He just kind of chewed, chewed over all the memories with humor and tears: times with his boys; parties; etc. This led to the third phase, the one covered in this book, creating sacred space for his inner work which sometimes included others like me and you, where we could all experience unconditional acceptance. This last phase was exploring the beauty and meaning of life. He could savor his existence. And he absolutely went far outside of everyday reality to complete his life and prepare for what's ahead. These three phases are somewhat nonlinear, overlapping and influencing each other.*

Stasche: *He found joy and connection in a lot of places on the way to the divine.*

Tom: *During* a *visit, as Stanley held my hand, he looked at it and asked, "Is this my hand." One way of looking at it is that he had lost his memory and perceptions and awareness. Another way of looking at it is that he had actually gained a larger universe by loosening the definition of his physical boundaries. That's what stunned me when I worked with him. It seemed he hadn't lost his identity; he had extended it to include me.*

Stasche: *We're not supposed to do these things in everyday life. Well, perhaps when falling in love, going crazy, or if you're on a spiritual quest experiencing the rapture of God.*

Why wait? Why wait for Alzheimer's dementia states of consciousness? Especially if you feel that they are going to happen to you sometime anyway. Why not enter these states consciously in your own ways or with the aid of a facilitator or with the guidance of the exercises we present in this book.

Tom: *I'm working on it.*

Stasche: *Yeah, me too. I try to follow my heart and stay in touch with other people and forces larger than myself. It's not easy all the time, though certainly rewarding. We stand on Dad's shoulders and the shoulders of that generation. Will we rise above the disease of Alzheimer's dementia onto a new level of inner awareness?*

Why did Dad have to wait for Alzheimer's to go into remote inner work states?

Tom: *There was a barrier to facing his grief and pain and guilt. There was a barrier for getting permission to go farther. Stanley needed permission and the "great permission giver" wasn't giving permission.*

Stasche: *So the great permission giver is a spiritual entity bigger than life but projected on and coming through his wife, his kids, organizations, society as a whole. Dad broke norms in so many ways with his humor and social activism, but he needed the Alzheimer's "boost" to go farther in his inner life. I feel like crying.*

End of life dreaming

Tom: *Also physical characteristics and behaviors at the end of life could differ from ordinary behavior. Things I may consider ordinary may be labeled dementia or Alzheimer's from the point of view of some caregivers. As a parallel, when I was growing up we would be romping outside before school, romping outside during lunch and recesses, and romping, wandering, and exploring all over the neighborhoods, woods, rivers, and construction sites after school until sundown. We would build model airplanes, fly 'em, crash 'em and then burn 'em or blow them up with cherry bombs. This behavior was seldom disturbing to the community and our caregivers in those days, but it is considered very disturbing to the community and caregivers these days. It is the point of view of the community and caregivers that has changed.*

I find it no surprise that people, toward the end of their lives on this planet, get the urge to wander, get agitated, etc. Granted, wandering and all the many accentuated behaviors associated with end of life are going to drive some caregivers crazy. But I prefer not to label spontaneous behavior in people only from the point of view of caregivers; just the way I prefer not to label the spontaneous behavior of kids only from the point of view of caregivers. I believe dropping personal history at the end of life, accompanied by unusual behaviors, is part of a perfectly normal stage, possibly a yearning for more detachment or a desire to work on resolving unfinished business. There may be need for new diagnoses, with less stigma attached, such as "end of life dreaming", in addition to Alzheimer's dementia.

Tom's dream

Tom: *So, I think of my dream from last week, where your dad's in a wheelchair. I'm pushing him in a wheelchair, in a race. There is a party at your house and you want to take a picture of your dad at the start of the race, and I'm pushing him in the wheelchair in the race. At first he's really animated and really excited about the race. And he lasts two laps, then I run out of stimulants for him. I can't keep him awake. He goes to sleep; he is remote, fishing trip, hunting trip, enjoying himself, far away, remote . . .*

Stasche: *Yeah, he did a lot of remote state things in his life, a lot of fishing that was a real outlet for meditation and going quietly into deep feeling states. We often went together. Hunting was quiet, with an acute visual and auditory awareness added. Both fishing and hunting required a deep intense waiting, for the fish bite or the bird flight.*

Dad had a great party nature and fun competitive personality, and in your dream I want to capture this on film, to preserve it.

Dad is in the race, but then, no matter how much you push, he won't be stimulated. After racing for awhile he has to go remote. He has another side. This book is a picture of that other side. His task the last seven years of his life was to withdraw and mostly do inner work.

Tom: *So why did he have to go extremely far away? Couldn't he have been more extreme within the context of normal reality?*

Stasche: Dad was definitely an "extreme" person at times with his humor and activism, by our culture's normal standards.

Tom: *Not extreme enough about bringing inner awareness out.*

Stasche: *He couldn't, and probably we can't be extreme enough around cultural prohibitions against altered states of consciousness. I hope I can find my inner self enough in our society to stay healthy, but I know that some ethical conflicts tear me apart. I hope I can maintain what Arny Mindell calls the two state ethic. I try to pay enough attention to the states of my outer identity and my inner identity and get them to communicate with each other as best I can, and suffer with the conflict between the inner and outer when I have to.*

Dad could have been even more "radical" in relating to people and in relating to the inner deep essence of his spirituality. And he did the best he could.

Alzheimer's career

Tom: *The preacher thing. He would have made a good preacher and might have gotten further in his spirituality.*

Stasche: *You and I have talked other times about Alzheimer's bringing enforced retirement, separation, and career change for Dad. Alzheimer's forced Dad to go beyond being the husband and provider, into his new spiritual inner work career. And what's interesting is that in doing the indexing for this book the three of the most common words are ecstasy, love, and death, certainly all spiritual aspects of his Alzheimer's career.*

Alzheimer's: beautiful eternity

Tom: *You know, besides the folks at Manteno, I spent the most time with your dad in his really remote Alzheimer's dementia universe. I think, if he could speak from there, he might say, "Alzheimer's is good for the soul!"*

Stasche: *Yeah, we do know for certain he chewed on "eternity" and in his own words, "It's BEAUTIFUL . . . it's as beautiful as it can be."*

Chapter Summary

1. **An estimated 26.6 million individuals** "suffer" from Alzheimer's dementia in the world today. If we were to locate them in the same geographic area, their country would rank as the 46th largest country in the world, larger than Saudi Arabia, North Korea, Australia, or Romania; and have the fastest growth rate. Within this country the principles of dementia would be the norm, and as one of the top 50 countries in the world, dementia would be one of the majority "normal" states in the world. Such statistics, coupled with Alzheimer's resistance to treatment, in our opinion, support our research hypothesis that Alzheimer's dementia is more than mere physical pathology, and ranks as one of many amazing modalities of consciousness in its own right, with its own psychosocial processes.

2. **Apprehension about people with Alzheimer's dementia:** Thanks to Ann Jacob for the following point. Early in the last century, many people with illnesses were marginalized. Now people with these same illnesses are treated with much more compassion and given support and encouragement. However, people with Alzheimer's dementia are often avoided and kept apart from larger society. And frequently, rather than being sympathized with, they are corrected: *Dad, you forgot such and such! Don't you remember?* Do we go around "correcting" people with cancer? A cultural attitude shift is taking place regarding people in Alzheimer's dementia states, a shift to more communication, compassion, and a deeper understanding of their essential natures.

3. **A deep soulful and spiritual growth process,** in our opinion, exists in the background of Alzheimer's dementia. Knowledge and awareness of this process offers hope and relief to patients, family members, and caregivers alike. The spiritual growth processes in the background of Alzheimer's dementia can be described as the following interbraided channels of the same river:

   a) working on "unfinished business", such as resolving individual and family issues, including meeting new aspects and partially lived aspects of the personality;

   b) "harvesting", such as recalling and savoring life experiences;

   c) "imparting blessings", such as openly accepting loved ones;

   d) creating "sacred space" for marginalized experiences by tapping into a sense of something larger than ourselves;

   e) garnering "meaning" by exploring formative experiences and essential beliefs;

   f) making "spiritual connections" like immersion in the beauty of eternity;

   g) Intimacy and deepening of relationships; and

   h) Working creatively on cultural and societal conflicts.

These processes are not linear stages of Alzheimer's dementia, but rather nonlinear descriptions of currents that may run separately or concurrently.

4. **For Alzheimer's dementia treatment** we advocate a multidisciplinary regimen which embraces medical and therapeutic interventions <u>and</u> addresses the underlying dis-ease by using Process Work or other therapies, techniques, or methods that can help facilitate inner sentient awareness in people experiencing Alzheimer's dementia. This awareness will benefit patients, families, caregivers, and society. (see p. 26-27)
5. **For Alzheimer's dementia "prevention"** we advocate a homeopathic approach to the dis-ease in concert with other appropriate medical and therapeutic treatments/techniques. "Homeopathic" in this sense means using small doses of the disease itself for prevention, the same approach as an inoculation. Here is our homeopathic "prescription" for "prevention":
   A. Each exercise in this book is designed to give a homeopathic dose of Alzheimer's. Do all the exercises now and repeat periodically in the future. Each time you do them you give yourself the opportunity to peel back one more layer, go one step deeper into your own deep inner sentient awareness, and the opportunity to help free yourself from the compulsory "call" of Alzheimer's.
   B. Create your own opportunities to explore the Alzheimer's state, such as visiting with those who have Alzheimer's dementia, or volunteering at an Alzheimer's dementia unit. Spend enough time during each visit to drop your own agenda, if any, and simply follow and experience the Alzheimer's dementia state in the moment and relate to people who are willing to relate with you.

C. Explore your own issues around physical, emotional, and spiritual intimacy and inner work through reading, meditation, prayer, counseling, therapy, Process Work, support groups, workshops, or whatever works for you.
D. Complete unfinished business and drop outmoded behaviors, beliefs, and personal history that no longer work for you.
E. Fully explore even minor episodes of forgetfulness instead of glossing over them.

In our next volume we will present research on people in early and intermediate stage Alzheimer's; the application of the above "prescriptions" to communicating with people in those stages; and the successes we have had, particularly with early Alzheimer's dementia, memory loss, and coma.

6. **Beauty:** As radical as it sounds to equate beauty with Alzheimer's dementia states, we discovered beauty both in our direct experience in relationship to Stanley, and in his own inner process, even in his most extreme emotional and remote Alzheimer's states. And we were startled to hear the word *BEAUTIFUL* in his clearly spoken description of his experience at the edge of death after awakening from a comatose state.
7. **This is a love story:** It is the story of the anguish and ecstasy of trying and succeeding, beyond all conventional wisdom, to stay in communication and in relationship with a loved one who is in and out of very remote states of consciousness, on his inner path. It is a story of human hearts searching, honoring, surrendering, celebrating, finding meaning, and embracing a process larger than ourselves.

8. **Families:** Stanley's end of life surprise party became a creative sacred ceremony which offered celebration, blessing, honor, and closure for Fran and Stanley, Stasche, Tom, family, and friends. This points to the potential for more families suffering from Alzheimer's dementia to experience this kind of deep intimacy and individualized ceremonial closure.
9. **Welcome:** This is our experience. We welcome and encourage your interest, your experiences, and healthy skepticism which will help develop the work.

Respectfully submitted, *Stasche and Tom*

Chapter 9 Exercise: Review, reflection, and application

1. Take a few minutes to review each of the exercises you tried throughout this book, and jot down at least a few words or ideas that come to mind. You may have skipped one or two exercises and want to go back and try them now before continuing to step 2:

   Chapter One (p. 22) ~ *Creative pre death celebration*

   Notes: ______________________________

   Chapter Four (p. 46) ~ *Ecstasy via Alzheimer's dementia*

   Notes: ______________________________

   Chapter Five (p. 84) ~ *Being Alzheimer's dementia*

   Notes: ______________________________

   Chapter Six (p. 111) ~ *"Dying"*

   Notes: ______________________________

   Chapter Seven (p. 134) ~ *Pacing the breath*

   Notes: ______________________________

2. Take a few minutes to reflect on what you have experienced from the exercises above. Then answer the following question:

   Has your knowledge or attitude or curiosity about

   Alzheimer's dementia changed, and if so, in what way?

   ____

   ____

   ____

3. How might your experience from this book benefit you?

   As a friend of someone with Alzheimer's dementia or friend of a caregiver:

   ___

   ___

   ___

   As a caregiver:

   ___

   ___

   ___

   As someone who might experience forgetfulness or dementia yourself, now or in the future:

   ___

   ___

   ___

4. What advice do you imagine offering to someone whose life is affected directly or indirectly by Alzheimer's dementia?

   ___

   ___

Bibliography and Resources

BOOKS AND ARTICLES

ANDREWS, Keith. "Misdiagnosis of the vegetative State: Retrospective Study in a Rehabilitative Unit." *British Medical Journal.* 1996;313;13-16 (6 July). BMJ Publishing Group. London. A study of 40 patients in vegetative state found 43% misdiagnosed. Most of the misdiagnosed were blind or severely visually impaired.

BARNES, Irene and TOMANDL, Stan. "Coma Care in End-stage Dementia" *Canadian Nursing Home Magazine.* White Rock, BC, Canada. V. 15. No. 1. March/April 2004. Special dementia care education issue.

Caregiver Guide: Tips for Caregivers of People with Alzheimer's Disease. National Institute on Aging Information Center. Gaithersburg, MD. 2005. ~ Available at www.nia.nih.gov

GROVES, Richard and KLAUSER, Henriette. *The American Book of the Dying: Lessons in Healing Spiritual Pain.* 2005. Celestial Arts. Berkeley, CA. ~ Ancient wisdom and practical real life experience for working with those near death. Contains a chapter on coma work.

MINDELL, Amy. *Coma, a Healing Journey: A Guide for Family, Friends, and Helpers.* 1998. Lao Tse Press. Portland, OR. ~ A practical guide to nonintrusive treatment of coma patients, especially for people with traumatic brain injury.

MINDELL, Arnold. *Coma: Key to Awaking.* 1989. Republished under: *Coma: The Dreambody near Death.* 1994. Currently available as e-book; call 503-222-3395. ~ This groundbreaking work with people in metabolic coma offers new directions in psychotherapy and in the study of people in near death states.

RICHARDS, Tom. *Eldership: A Celebration.* 2006. Interactive Media. Glenview, IL, USA. ~ www.lulu.com/sentientcare "One of my fondest dreams is that we may rediscover, support and celebrate the blessings, inspiration and treasures of eldership in ourselves, our families, our workplace, and our communities."

***The Rush Manual for Caregiver.* Rush Alzheimer's Disease Center. Chicago. 2002. ~ A comprehensive general information manual for family members and professional caregivers. Available online: www.rush.edu/patients/radc/**

TOMANDL, Stan. *Coma and Remote State Directive.* 2005. Coma Communication. Victoria, BC. www..lulu.com/sentientcare ~ A living will for those concerned with communication and decision making during states of confusion, delirium, stupor, coma, vegetative state, depression, catatonia, dementia, and other remote states of altered consciousness

TOMANDL, Stan. *Coma Work and Palliative Care: An Introductory Skills Manual for People Living in Delirium and Coma.* 1991. Coma Communication. Victoria, BC. www.comacommunication.com ~ Bedside manual containing detailed techniques for working with people in states of altered consciousness, delirium, advanced dementia, and metabolic coma.

VALEO, Tom. "How to Treat Ailing Memory" *Neurology Now, Mar/Apr 2006.* ~ Alzheimer's evolves over decades, interacting with many other factors, showing up first as mild cognitive impairment.

VICTORIA HOSPICE SOCIETY. *Hospice Resource Manual: Volume I; Medical Care of the Dying.* 1990. Victoria Hospice Society. Victoria, BC. ~ An incredibly detailed manual, including pre death biological and psychological changes in people.

ZIMMER, Carl. "What if There Is Something Going on in There?" *New York Times.* October 02, 2003. New York. ~ Functional MRI research at the Sloan Kettering Institute by Nicholas D. Schiff and Joy Hirsch demonstrating that some people in vegetative state have normal brain function when given meaningful stimuli.

PROCESS ORIENTED COMA WORK WEB LINKS AND EMAIL ADDRESSES

Coma Communication and Process Oriented Facilitation ~ Canada & USA ~ Stan Tomandl and Ann Jacob's website
www.comacommunication.com/

Memory Enhancement ~ Tom Richard's website ~ USA

www.tomrichards.com/alzheimers.htm

COMAcare: Hearing the Silent Voice ~ South Africa
www.comacare.com/

Drs. Arnold and Amy Mindell ~ USA

www.aamindell.net/coma.htm

Process Work Institute ~ USA

www.processwork.org/Coma.htm look under Coma

Dr. Pierre Morin ~ USA

www.creativehealing.org/coma.htm

Gary Reiss, PhD ~ USA

www.garyreiss.com/coma.htm

Max and Ellen Schupbach ~ Germany

www.grossetransformationen.de

Ingrid Rose at 503.248.1608 ~ Portland, Oregon, USA
email: ingridrose8@cs.com

Lily Vassiliou at 30-694-220-4957 ~ Athens, Greece
email: lvassiliou@gmail.com

AGING

Pacific Institute: counseling, education, research
http://www.pacificinstitute.org/index.html

AgeSong Senior Communities
http://www.agesong.com/home.htm

Center on Aging, University of Victoria
http://www.coag.uvic.ca

National Institute on Aging Information Center
http://www.nia.nih.gov

ALZHEIMER'S AND OTHER DEMENTIAS WEB LINKS

Alzheimer's Research Foundation ~ USA
http://www.alzinfo.org/

Alzheimer's Society ~ UK
http://www.alzheimers.org.uk/

Alzheimer Society of Canada
http://www.alzheimer.ca/

Alzheimer's Association ~ USA
http://www.alz.org/

Alzheimer Research Forum ~ USA
http://www.alzforum.org/home.asp

Alzheimer's Disease Education and Referral Center ~ USA
http://www.alzheimers.org/

Buckner, Dr. Randy
http://www.hhmi.org/news/buckner5.html

HOSPICE AND PALLIATIVE CARE WEB LINKS

International Association for Hospice and Palliative Care
http://www.hospicecare.com/

Hospice Africa
http://www.hospiceafrica.org/

Foundation for Hospices in Sub-Saharan Africa
http://www.fhssa.org/i4a/pages/index.cfm?pageid=1

Canadian Hospice and Palliative Care Association
http://www.chpca.net/home.htm

European Association for Palliative Care
http://www.eapcnet.org/

American Association of Hospice and Palliative Medicine ~ USA
http://www.aahpm.org/

National Hospice and Palliative Care Organization ~ USA
http://www.nhpco.org/templates/1/homepage.cfm

Hospice and Palliative Care Nurses Association - USA
http://www.hpna.org/

National Association of Home Care ~ USA
http://www.nahc.org/

Hospice Education Institute ~ USA
http://www.hospiceworld.org/

Center to Improve Care of the dying ~ USA
http://www.gwu.edu/~cicd/

International Observatory on End of Life Care
http://www.eolc-observatory.net/global_analysis/index.htm

American Board of Hospice and Palliative Medicine ~ USA
http://www.abhp.org/

SPIRITUAL CARE WEB LINKS

Sacred Art of Dying ~ USA
http://www.sacredartofliving.org/programs/sad.htm

The Blessed Foundation ~ Cleveland, OH, USA
http://www.blessed-foundation.com/pages/5/index.htm

The Listen Centre ~ UK

http://www.the listencentre.com/

TRAUMATIC BRAIN INJURY WEB LINKS

Coma Recovery Association ~ USA
http://www.comarecovery.org/

Medical protocol for the prevention of vegetative state ~ USA
http://www.pbs.org/wgbh/nova/coma/

Traumatic Brain Injury Survival Guide
http://www.tbiguide.com

National Institute of Neurological Disorders and Stroke
http://www.ninds.nih.gov/disorders/tbi/tbi.htm

Brain Trauma Foundation
http://www.braintrauma.org/

VEGITATIVE STATE WEB LINKS

National Institute of Neurological Disorders and Stroke ~ USA
http://www.ninds.nih.gov/disorders/coma/coma.htm

HealthLink – Medical College of Wisconsin ~ USA
http://healthlink.mcw.edu/article/921394859.html

Persistent Vegetative State
http://www.thalidomide.ca/gwolbring/pvsilm.htm

Persistent Vegetative State
http://www.cwu.edu/~chem/courses/Chem564/finalpapers/PVSfinal.html

Index

Tom Richards evolved from a corporate executive (BS in electrical engineering, Cornell University; MBA in marketing, University of Chicago; and CPA, University of Illinois) to a professional educator, researcher, and consultant working with health issues and people in altered consciousness. He has studied, researched, and applied Process Work for the past twenty years. Tom uses his awareness abilities to follow people's processes at their deepest sentient levels, encouraging beauty and eldership to come forward, even at seemingly impossible times. He works with individuals experiencing body symptoms, injuries, illness, surgery; and memory loss ranging from mild cognitive impairment to Alzheimer's and other dementias, end of life processes, and coma.

Stan Tomandl, MA, PWD, Process Work Diplomate has a passion for all of nature giving him a heart connection with the essence of human cycles, especially the sacredness of those living in altered consciousness. For over 20 years, Stan has specialized in working with people in near death states, memory loss, dementia, traumatic brain injury, and coma, along with everyday life issues. He is an instructor and lifelong learner at the Process Work Institute of Portland, Oregon, author of *Coma Work and Palliative Care,* and contributing editor to *The American Book of Dying.* Stan, cofounder of Coma Communication, serves on the faculty of the Sacred Art of Living Center, and the board of ComaCARE, South Africa with his wife Ann Jacob. They train, coach, and supervise caregivers, chaplains, family members, volunteers, and educators.

LONG BEFORE THE ALZHEIMER'S JOURNEY

Stanley Tomandl in his stylish hat holding baby Stan (Stasche), with his wife Fran, center on the hood of their Packard automobile, at a family picnic in 1946.

Contact information:

Tom Richards: www.sentientcare.com
Stan Tomandl: www.comacommunication.com